Seth Pancoast

What is Bright's disease?

Its Curability

Seth Pancoast

What is Bright's disease?
Its Curability

ISBN/EAN: 9783337155926

Printed in Europe, USA, Canada, Australia, Japan

Cover: Foto ©berggeist007 / pixelio.de

More available books at **www.hansebooks.com**

WHAT IS

BRIGHT'S DISEASE?

Its Curability.

WHAT IS

BRIGHT'S DISEASE?

Its Curability.

By SETH PANCOAST, M. D.,

Specialist in the Treatment of Chronic Diseases; Author of Ladies' Medical Guide, Treatise on Consumption, The Kabbala, etc., etc.

WITH ILLUSTRATIONS.

PHILADELPHIA:
PUBLISHED BY THE AUTHOR.
1882.

TO

THE SUFFERERS

FROM

BRIGHT'S DISEASE,

WITH

A HOPE THAT THEY MAY FIND INSTRUCTION

AND

CONSOLATION IN ITS PERUSAL.

PREFACE.

It is customary in the preface for the author to assign reasons, and offer an apology, for writing the work. We have reasons for writing which we will state, but no apology to offer.

The disease of which we treat in this work is regarded hy many physicians as incurable, even in its acute, but especially in its chronic form. In our opiuion, the failure to treat it successfully is due largely to a wrong conception of its nature and cause, and this book is written for the purpose of expressing the author's ideas, and affording hope to those suffering from this terrible malady. We claim that the primary cause lies in the organic nervous system which controls the nutrition and growth of the entire organism, as well as the elimination of the products of disintegration.

In the secreting structure of the kidneys the disease is supposed to be first manifested. This is a mistake, for it may exist for many months, if not years, before albumen is detected in the urine. The kidneys are not the only organ involved when the disease becomes fully developed; we find the heart, lungs, and liver participating, and this cannot be from sympathy with the kidneys, but is due to an ennervation of the NERVO-VITAL energy, the master-workman in controlling the functions of the physical organism. The conditions under which Bright's Disease makes its appearance, may be lurking in the organic nervous system for years, before the patient, his friends, or even his physician is cognizant of it, and all at once, some exciting cause, such as mental strain, cold, debauch, etc., reducing the vital energy, brings it to the surface. So long as it is treated as a local disease, exclusively, successful results cannot be expected.

We have fully demonstrated its curability within the last few years, to our own satisfaction, as well as to that of many we have treated. The ennervation of the vital energy centered in the organic nervous system must be corrected or replaced by normal action before local treatment applied to the kidneys can assist in reducing the inflammation, thus enabling the vital energy to overcome structural changes that have occurred, and to restore the organ to its normal action. Great care must be observed by the patient for fear of a relapse, which is liable to occur until *normal vital action is firmly re-established.* Our motive in writing this work is evident: it is to set forth our views on Bright's Disease, and to assure those who are suffering from it of its curability, if there be not already too great an ennervation of the vital energy and disintegration of the kidneys.

In conclusion, we would remark that repetition of words and phrases, and even of statements, is unavoidable in a work of this character, written expressly for non-professional readers.

SETH PANCOAST, M. D.

917 ARCH STREET,
March 1st, 1882.

CONTENTS.

CHAPTER I.
THE STRUCTURE OF THE KIDNEYS, ... 15

CHAPTER II.
THE NERVOUS SYSTEM, ... 33

CHAPTER III.
FUNCTION OR OFFICE OF THE KIDNEYS, ... 55

CHAPTER IV.
ABNORMAL CONDITION OF THE KIDNEYS IN BRIGHT'S DISEASE, ... 83

CHAPTER V.
WHAT IS BRIGHT'S DISEASE? IT'S CURABILITY, . 97

CHAPTER VI.
SYMPTOMS OF BRIGHT'S DISEASE, ... 116

CHAPTER VII.
CAUSES OF BRIGHT'S DISEASE, ...

CHAPTER VIII.
ADVICE TO THOSE SUFFERING FROM BRIGHT'S DISEASE, ...

DEFINITIONS.

CORRELATION AND ITS COGNATES.—"Correlation," is derived from the Latin *correlatio*, a triple-compound word, from *latus* the participle passive preterite of *fero*, with the successive prefixes *re-*, and *con-*; etymologically, it denotes "a bearing back upon" a person or thing; thus, the ordinary dictionary definition is "a reciprocal relation," etc.; but we employ this substantive, "correlation," and its cognate adjective and verb, in a specific or technical application, to denote the peculiar relation that obtains between the Subjective, Primary Force, on the one hand, and the Objective, Secondary Force, on the other hand. The former is the direct instrument of the Spirit, and is the Vital, the Life Force; the latter is the same manifested in material organization, as Nerve Force—the one centres in the Soul, the other in the Nervous System of the Living Organism. As the Soul and body of every living entity are developed simultaneously, the Primary and Secondary Forces must *correlate* in action and operation.

NERVO-VITAL.—The foregoing definition suggests the signification of this term; the Spiritual Vital Force of the Soul, to develop and sustain the physical body, must establish itself with the Secondary Nerve Force, and the two correlate as the Nervo-Vital Force.

CHAPTER I.

STRUCTURE OF THE KIDNEYS.

[See Frontispiece.]

THE form of the kidneys somewhat resembles the ordinary bean, the inner margin being concave and extending inward towards the spine; the outer edge is thick and rounded and directed outwards. They are situated deep in the lumbar region, on each side of the spine; the right is slightly lower than the left, being depressed by the liver which lies immediately above it. The kidneys are occasionally out of their natural position—either in front of the spine or much lower down in the abdominal cavity —even as far as the cavity of the pelvis. They are embedded in adipose or fatty tissue, which is very profuse in fat persons, while in thin, spare individuals, there is scarcely a trace of it. The average length of these organs is four to four and a half inches, breadth two inches, and thickness one inch; the weight in males is five to six ounces, in females, four to five ounces.

They are partially covered by the peritoneum, which lines the abdominal cavity. The right kidney is covered by the ascending colon or the ascending part of the large intestine, and the left by the descending colon, or descending part of the large intestine. The right kidney is also in contact with the duodenum and covered by the right lobe of the liver, and occasionally a part of its anterior surface is covered by the gall-bladder. The left kidney lies in contact with the spleen and is covered by the large end of the stomach, when it is distended during the process of digestion. It is important to understand the relation of the kidneys with the colon or large intestines. Abscess of the kidneys has been known to burst into the colon, and the ulceration of the intestines would cause serious structural change in the kidneys.

On the concave and spinal margin of the kidney is a small depression or fissure, called the hilum, through which the arteries and nerves pass into the kidney and the veins pass out. It is this depression where the pelvis of the kidney begins. The ureter or excretory duct extends from the hilum to the base of the bladder. It is a small canal composed of elastic tissue, varying in size from a chicken to a goose quill. At its commencement in the kidney and its terminus in the bladder, it is dilated but becomes narrowed near the middle. It is in the contracted portion of the ureter that urinary calculi or gravel become lodged, in their passage from the kidneys to the bladder, causing excruciating pain. This tube is lined by a delicate mucous membrane similar to that which lines the bladder and kidneys.

THE BLOOD-VESSELS OF THE KIDNEYS:— These consist of the *renal* artery and vein; the former is a branch of the abdominal aorta and, previous to its passage into the kidney, it gives off branches to the capsule, ureter and surrounding cellular tissue. After it enters the kidney, it divides into four or five branches and again subdivides into smaller branches forming the *Malpighian tufts*, which we will describe when speaking of the minute structure of the kidneys.

The vein rises in the substance of the kidneys from the capillaries and conveys the blood out of the kidneys into the *vena cava*.

2*

18   BRIGHT'S DISEASE.

THE NERVES OF THE KIDNEYS:—These consist of branches from the lower and outer border of the *semi-lunar ganglion* and the

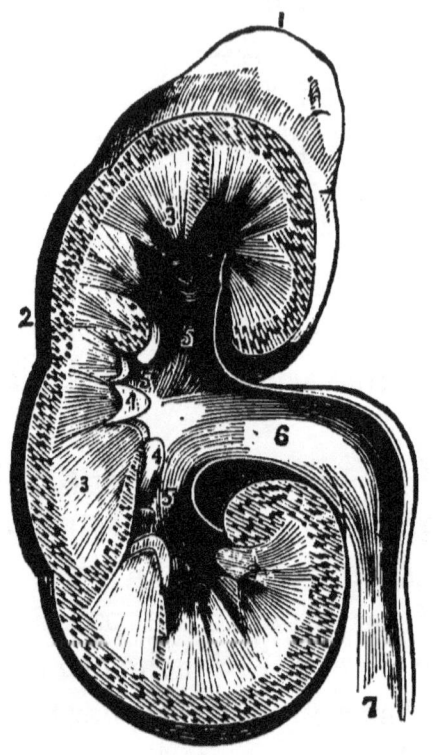

FIG. 1.

Longitudinal section of the kidney, with its renal capsule. 1, Renal capsule; 2, cortical or vascular parts of the kidney; 3. 3, uriniferous tubes collected into a conical form; 4, 4, papillæ projecting into their corresponding calces; 5, 5, 5, the three infundibula; 6, pelvis of the kidney; 7, ureter. (After Dr. Morton.)

solar plexus formed by the descending branches of the small *splanchnic* nerve.

THE GENERAL STRUCTURE OF THE KID-

NEYS.—If a kidney be moved from its position and stripped of its adipose and loose fibrous tissue, we find it completely covered with a capsule, formed of dense fibrous tissue, which not only encloses the body of the kidney but is continuous with the pelvis and ureter. If a longitudinal incision be made through its outer border, we perceive that it is composed of two distinct portions differing materially in appearance. The external is called the *cortical* and the internal the *medullary* portion; the former is about two lines (one-sixth of an inch) in thickness. Its color will depend upon the quantity and character of the blood it contains. When normally supplied with healthy blood, it is of a light-red color. In anæmic subjects where there is a deficiency of red corpuscles, it presents a yellowish-white appearance. If we examine the structure with a low magnifying power, we find it made up of secreting tubes called *tubuli uriniferi*, blood vessels, and *Malpighian tufts* or *glomerules*. In the cortical substance these tubules are very tortuous; as we descend in the substance of the kidney, they become straight, forming the pyramids of Malpighi.

The *medullary* substance, or as it is sometimes called the tubular portion, of the kidneys is much less dense than the cortical substance and assumes the form of cones or pyramids with the *apices* downward, and the bases upward. The *apices* of these pyramids are called *mammillary processes* or *papillæ*. The number in each kidney varies; usually, however, there are from twelve to fifteen in each. Some of them are compound, being formed by the union of two which have a common mammillary termination. The cut surface of a pyramid presents a striated appearance, being composed of tubes, named the tubes of Bellini, who first discovered their tubular character. They are united by a fine net-work of fibrous tissue, in which we find associated some large veins which assume a straight course between the tubes. The papillæ of each of the cones have several openings, from which the urine escapes into the *calces* and *infundibula*. No Malpighian bodies are found in the medullary portion of the kidneys; they only exist in the cortical substance.

PELVIS OF THE KIDNEY.—This is a strong,

PELVIS OF THE KIDNEY.

fibrous pouch, divided into three compartments, called *calces*—one at each end, and one intermediate or in the middle. Each

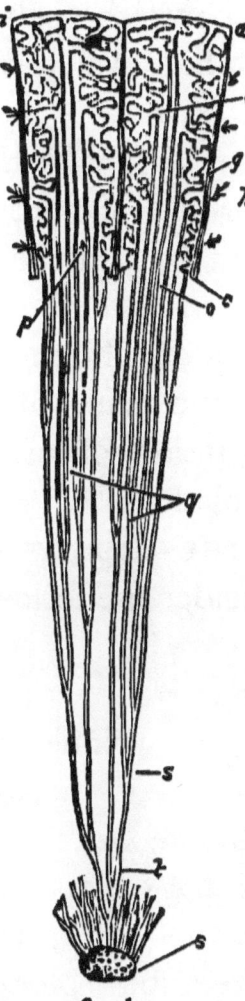

FIG. 2.

Vertical section from the cortical surface of the kidney to the apex of one of the medullary cones q, c, bundles of straight tubes passing towards the surface, a, i; the same observed at o and p; at l, the straight tubes terminate in what is designated Henle's loop and become tortuous; q, s, t, straight tubes with their dichotomous divisions; h, artery terminating in a cluster of Malpighian bodies; s, apex of one of the medullary cones terminating in a mammillary process or papilla. (After Schumlausky.)

one of these calces is again divided into three or four funnel-shaped processes, called *infundibula*, which embrace the papillæ at their upper or basal extremity. In some anatomical works *calces* and *infundibula* are synonymous or convertible terms. From these calces the urine passes into the pelvis of the kidney, which is of a flat, oval form, terminating in the ureter.

THE URETER.—The ureter is about the size of an ordinary quill, formed of fibrous tissue, lined its entire length by a mucous membrane. It is from fifteen to eighteen inches long, and enters the bladder obliquely, which obliquity answers the part of a valve, preventing a regurgitation of urine after it has passed into the bladder.

MINUTE STRUCTURE OF THE KIDNEYS.—In this division we will consider the following structures:

1. Fibro-cellular structure.
2. The *Tubuli Uriniferi*, or uriniferous tubes.
3. *Malpighian Tufts*, or *Glomerules*.

1. *The Fibro-cellular Structure.*—This portion consists of a fibrous net-work, the meshes of which support the tubes, blood-vessels, nerves, etc., and is therefore the frame-work of the kidneys. It consists of fibro-cellular tissue, and its office is mechan-

ical; it undergoes no change, and is therefore regarded as permanent tissue.

2. *The Uriniferous Tubes.*—It is diffi-

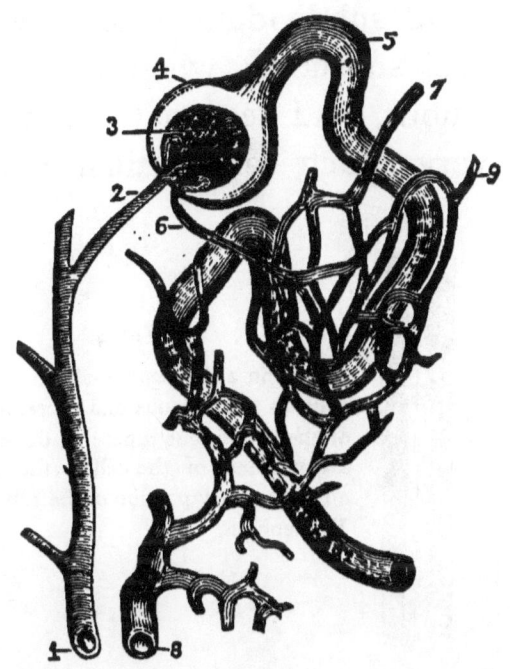

Fig. 3.
Microscopic view of secreting structure of the kidney. 1, the renal artery, which at 2 sends a twig to the Malpighian copuscle; 3, convoluted tuft, capillary ball or glomerule, enclosed by the capsule. The capsule, is seen at 4, and at 5 contracts into a tortuous uriniferous tube or duct of Terrian; 6, the efferent vein bringing back the effete blood, joins the veins 7 and 9 from other capsules, and thus forms the venous capillary plexus around the uriniferous tubes; 8, these capillaries converge and end in the emulgent vein. (After Dr. Morton.)

cult to trace these tubes from either the cortical substance, or from the apex of the medullary cone. If we commence at the latter point, we find them pursuing a comparatively straight course in the medullary portion of the kidneys [Figs. 1 and 2], but dividing and subdividing until they reach the cortical substance, when they become very tortuous, and terminate in a closed cavity, very much larger than the tube

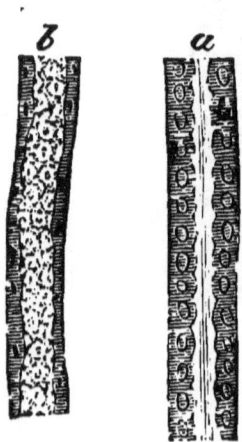

FIG. 4.

a, The arrangement of the epithelium cells in the tortuous and excretory portion of the uriniferous tubes; *b*, the epithelium arrangement of the cells in the straight or non-excretory portion of the tubes. (After Heidenhain.)

[see Fig 3]. It is the division and subdivision of these tubes that forms the base of the pyramid, and causes it to be uppermost. The spherical enlargement of the tubes is called the capsule, and is 1-150th of an inch in diameter [see Fig. 3]. The tube below the capsular neck is 1-480th of an inch in diameter. Their structure is very delicate, consisting of a basement or germinal membrane, and an epithelium layer of cells. As it is important to make our subject clear to the unprofessional reader, it will be necessary for us to be very exact in our description of this germinal membrane, and the mucous cells that rest upon it [Figs. 4 and 5]. This structure was first dis-

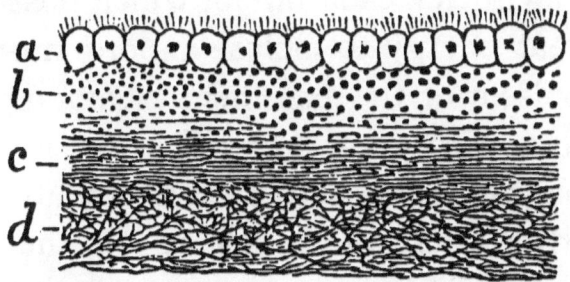

FIG. 5.
Diagrammatic representation of the basement membrane, and the vital changes that occur in its protoplasma, in the development of epithelium cells ; *d*, capillary blood-vessels which form the floor of the basement membrane; *c*, the protoplasma eliminated from the blood-vessels, called the basement membrane or formless matrix; *b*, the same vitalized by nervo-vital energy, becomes granular, which aggregate and form nuclei; *a*, epithelium layer of cells resting upon the outer surface of the so-called basement membrane.

covered by Bowman and Professor Goodsir; the former named it the *basement* membrane, on account of its being the basis upon which the epithelium and epidermic cells rested, and the latter called it the *primary* or *germinal* membrane, believing it to furnish the germs of the above named cells. In its most primitive form it is a homogeneous fluid layer, so minute that it is incapable of measurement. In its more advanced state of development, we find on the upper surface granular particles, upon which rest the epithelium cells. Before these cells are formed the granules unite, forming nuclei, each one of which possesses inherent energy to inclose a portion of the protoplasm of the basement membrane, by the formation of a cell-wall. The nucleus is now enclosed, and is in a condition to impart vito-chemical change to the cell contents. This consists of converting the contents of the cells resting upon the basement membrane into the solid constituents of the urine, which is a purely vital process. When this metamorphosis occurs, the life of the cell ceases, and its contents pass into the uriniferous tubes; other cells immediately form to take their place. The integration and disintegration of epithelium cells is very rapid, depending, however, upon the amount of fluid blood iliminated, and the vital energy imparted by the organic or sympathetic nervous system. We will show in a future chapter that the office of these cells is to eliminate the morbid products of decay containing nitrogen. In Bright's Disease the energy imparted by the organic nervous system is impaired, causing a change of function, and if the disease is not arrested, a structural change in the tubes takes place, and finally, completely suspending the office of the kidneys, causes a retention in the blood of nitrogenous products, blood poisoning, and death. We thus perceive that the basement membrane is an exudation from the blood, to which the organic nerve filaments are distributed, changing it to a granular form, which subsequently unite and form nuclei. The formation and character of the epithelium are the same in all glandular structures, yet their functions differ. In the liver they secrete bile, consisting of carbon and hydrogen; in the kidneys, urea, etc., containing nitrogen; in the stomach, a gastric juice; in the glands of the mouth, a salivary secretion, etc.

Attached to the outer surface of the epithelium cells are hair-like appendages, called cilia [Fig 5]. They vary from 1-500th to 1-1000th of an inch in diameter, and when in motion bend from their attachments in the cell to their points, returning again to their original state, like a field of wheat when depressed by the wind. Their motion is derived from the nucleus of the cell, and their office is to assist the movements of the products secreted by the cells, their motion being downwards. In some organs, particularly the reproductive of the female, their movements are upwards and inwards. Ciliary motion continues as long as the cell retains its vitality, the motion continuing sometimes after death has apparently occurred. It is stated by some writers to continue until decomposition ensues.

3. *Malpighian Tufts*, or *Glomerules*.—These bodies have been objects of great interest since their discovery by the distinguished anatomist whose name they bear [Fig. 3 and 6]. Malpighi believed them to be internal glands, which could be readily injected from the arterial branches to which they were appended. He also believed the urinary constituents were separated from the arteries of these bodies, and that the uriniferous tubes were excretory ducts of these glands. Schumlausky was the first to arrive at a correct knowledge of

these bodies and their connection with the uriniferous tubes. He described the Malpighian bodies as consisting of *glomerules* of vessels connected on one side with the

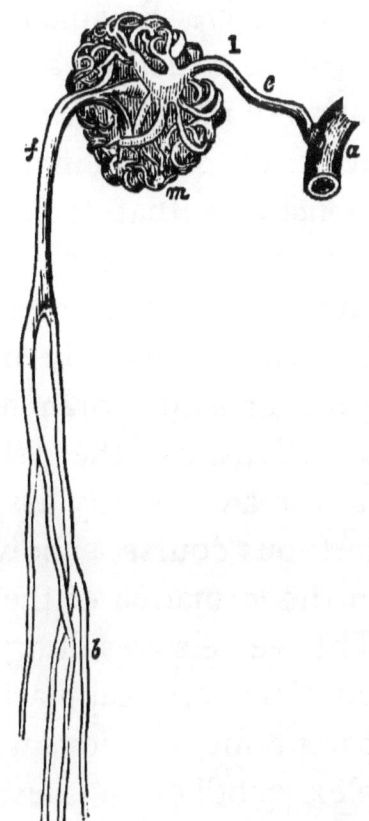

Fig. 6.
Malpighian body, etc., from the horse. *a*, arterial branch; *l*, the arterial twig, or afferent vessel; *m*, the Malpighian tufts; *f*, the vein, or afferent vessel; at *b*, its branches enter the medullary cone. (After Bowman).

arteries, and on the other side with the veins. He further believed there was a close connection between them and the tubes, but was unable to define it. From 1788, the time of Schumlausky's work, to

1842, when Mr. Bowman published his paper on the same subject, but little had been added to the knowledge of the Malpighian bodies.

We give Bowman's description of this intricate structure, and the wonderful adaptation of means to ends, affording undoubted evidence of an infinite Designer. He demonstrates that the capsule on the upper extremity of the uriniferous tubes is pierced by a twig of the renal artery [Fig. 3], which suddenly breaks up into three, four, or eight branches, diverging in all directions, like the petals on the stalk of a flower, and usually assuming a more or less tortuous course, subdividing once, or more, in the formation of the vascular ball or tuft. The vessels resulting from these various divisions are capillaries, with thin, transparent, membranous walls. After this complex subdivision they twist into a single, small vessel about the size of the arterial twig from which the ball originated. This vessel then emerges from the capsule, near the artery. Both twigs, as they pass in and out, adhere closely to the capsule. This tuft or capillary ball is held together by a mutual interlacement, there being no other tissue admitted into the capsule; it lies perfectly loose and bare in its inclosure. Since Bowman's discovery the German and French physiologists particularly have asserted that the glomerules are lined by a delicate mucous membrane, which is the first to undergo a change in Bright's Disease. Be this as it may, the office of the epithelium must be unnecessary for the elimination of the watery portion of urine, which is merely a filtration.

THE CIRCULATION OF THE KIDNEYS.—It is important to understand the circulation of the kidneys, in order to familiarize ourselves with their functions, in health and in disease. We have previously stated that the renal artery sends a twig to the capsule of the uriniferous tubes which divide and subdivide, forming a tuft of capillaries. These terminate in a single vessel which passes out of the capsule and anastomoses with branches from other Malpighian bodies, and they form a plexus of capillaries which anastomose around the uriniferous tubes, forming a continuous net-work outside the tubes and in close contact with its basement membrane. We thus have two sets of capillary vessels in the kidneys, one forming the glomerules, the other the plexus around the tubes [Fig. 3]. This plexus converges and ends in the emulgent or renal vein [Fig. 6] which emerges from the fissure of the kidney and terminates in the *vena cava*. It is well to note, that arteries convey blood *to* an organ which terminate in capillaries, and the capillaries terminate in veins which convey the blood *from* an organ. The arteries take blood to tissues and organs to supply them with nutrition and the veins take it back to the heart, leaving the products of waste at the excretory organs. When speaking of the function of the kidneys, we will show that the uriniferous tubes secrete the solid constituents of the urine, and that the Malpighian tufts eliminate the watery portions. The nerves of the kidneys will be explained in a separate chapter in which we shall speak of the vital energy.

CHAPTER II.

The Nervous System.

As it is in the nervous system that the vital energy is located, it will not be out of place to devote a short chapter to its structure and functions. All organic structure in which *active life* exists, has a nervous system of some form; it may consist of simple granules in a mass of protoplasm, as we find it in the *amœba*. Some deny its existence in the lowest organic forms, which are apparently structureless, yet there is evidence of sensibility and automatic action, and as these are only to be found in the nervous system, the inference is that such a system exists in these lower organisms, in a very rudimentary form. The granules are centres of vital energy, which subsequently become aggregated in nuclei, having the power of forming cells out of protoplasm. Therefore the most rudimentary form of nervous system consists of granules or corpuscles, possessing a centralizing energy, capable of manifesting motion in unorganized or structureless protoplasm. As the evolution of forms takes place, the granules aggregate and form nuclei, in which we have the energy increased. In the higher state of organic unfoldment we find the nuclei enclosed by cell-walls, which they have the power of forming. In order to concentrate the energy to meet the requirements of higher organic development, ganglia are formed, which are the aggregation of cells, free nuclei, granular particles, albumen, and fatty matter.

In the *Radiata*, we find a single ganglion with projecting nerve fibres; in the *Articulata* the form is more complex, and a greater differentiation is required. There are several ganglia joined together by nerves called commissures, and nerves leading to and from the ganglia to different parts of the organism. In the lower types of *Vertebrata*, a step higher in the evolution of form, we have a spinal system, and a miniature brain, which is gradually unfolded in the successively higher types until perfected in man. We thus perceive that a centre of nervo-vital energy may be a simple granule, a nucleus, or a ganglion.

THE NERVOUS SYSTEM.

The nerves are entirely different in structure from the ganglia, and possess a very different function. The ganglia are centres, and the nerves are the distributers of energy. A nerve is a minute tube, in the centre of which is a medullary substance, consisting of albumen and fatty matter. Embedded in this pulpy mass that fills the entire tube, is a delicate fibre, which is the nerve proper, and called the *axis cylinder;* this central fibre is uniform and continuous, unlike the pulpy mass that surrounds it. Inclosing the medullary mass is a dense, fibrous tissue that forms the wall of the tube called the neurilemma, which is continuous with the ganglia, and forms their outer structure. At the peripheral extremity of the nerve the medullary ceases [Fig. 7], in order to expose the nerve fibre to the granular mass, to receive impressions from the external world, and convey them to the centres of consciousness. After the fibre leaves its medullary encasement it forms into an anastomosing loop and returns again to its pulpy covering. When a nerve enters a ganglion it also leaves its medullary envelope, and when it emerges it again receives it, showing conclusively that it is the central fibre that is the carrier of energy from the surface of the body to the seat of consciousness and that receives energy from the ganglia and conveys it to different parts of the

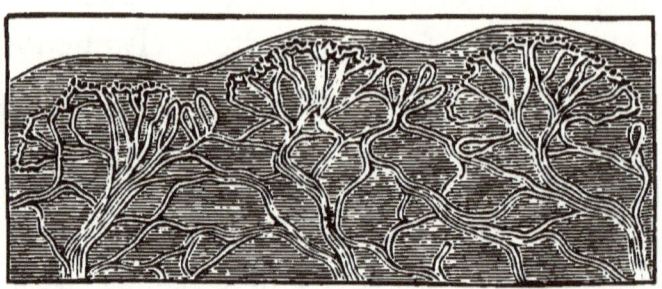

Fig. 7.

Looped termination of nerve fibres in the papillæ of the skin, with the granules in which the *central nerve fibre* communicates.

system. We thus perceive that the fibre unrobes itself, as it were, to receive energy; when it has done so, it robes itself to retain its energy until it is conveyed to where it is required. The medullary structure completely insulates the fibre, as the cable is insulated that lies at the bottom of the ocean. There is a remarkable similarity between the electro-magnetic battery and its insulated wires and the ganglion and its nerves [Fig. 8]. The ganglion contains cells and nuclei in which is generated nerve-force, and the nerves distribute this force,

THE NERVOUS SYSTEM.

the same as the electro-magnetic battery generates electro-magnetic force and the wires distribute it. These nerves are distributed to all parts of the body. This cannot be better illustrated than by comparing it to a fine spider's web, the meshes of which

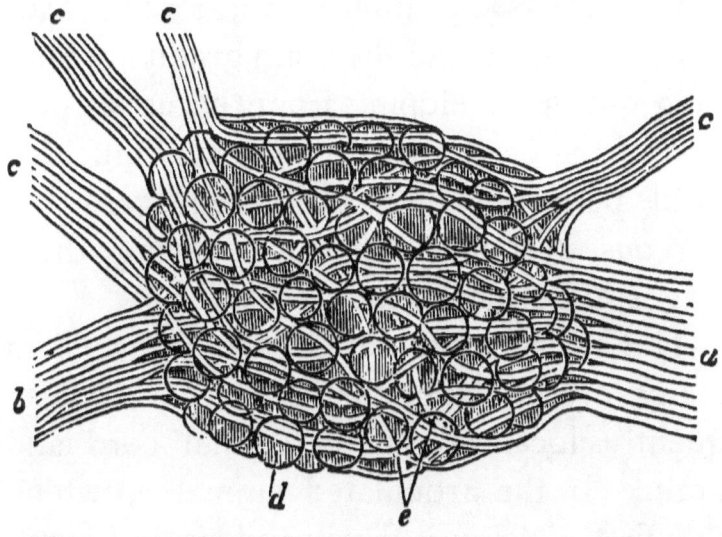

FIG. 8.

Ganglion of the organic or sympathetic nervous system of the mouse. *a, b*, nerve fibres connecting the adjacent organic ganglia; *c, c, c, c*, nerve branches to the viscera and spinal nerves; *d*, ganglionic cells, the seat of nervo-vital energy; *e*, nerve fibres passing through the ganglion receiving energy from the cells.

are so small that a needle's point could not touch any portion of the soft tissues without injuring one of these nerve fibres.

The nervous system is divided into the cerebro-spinal (consisting of the brain and spinal cord), and sympathetic or organic nervous system. The latter is the first to be developed, appearing in the lowest organic forms and is perfected in the articulated animals. In the vertebrated animals, which follow after the articulated, the spinal system makes its appearance, and, as we advance in this class, we find the brain gradually budding out or developing from the upper portion of the cord, which is not fully unfolded until perfected in man. We regard the nervous system as a unit, having three stages of development.

1. The sympathetic in which we find many of the functions, which are subsequently localized in the spinal cord and brain. In the articulated animals in which this first division is developed we find sensibility, and automatic action, consciousness, muscular co-ordination, emotions, the sexual passions and instinct, all of which are transferred to the spinal cord and brain when they are separated from the sympathetic nervous system. When the spinal cord is developed, automatic action is transferred to the medulla oblongata, the upper portion of the spinal cord. When the base of the

brain is unfolded, the seat of consciousness is transferred there. When the cerebellum or posterior of the brain is developed, the emotions, the sexual passions and muscular co-ordination are transferred to it. When the cerebrum or anterior brain is unfolded, we find the mind, which in the lower order of animals is represented by instinct. We thus perceive a gradual differentiation of the nervous system from the lower organic forms to man. These systems are so intimately connected that a serious injury occurring in any one of them finds a sympathy in the others.

Notwithstanding the separate functions assigned to each system, there exists a reciprocal relation; for example, if the pneumo-gastric or par-vagum nerve, which controls the action of the heart and respiration, be divided, the sympathetic nervous system cannot perform its normal functions, for the blood ceases to circulate, and oxygen, which is essential for chemical change, is not introduced into the blood by the lungs. Therefore nutrition, which office is assigned to the organic or sympathetic nervous system, cannot be carried on. If the sympathetic or organic system should be so impaired that it cannot perform its office, which is the elimination of the products of waste, through the excretory organs and of nutrition, the spinal cord and brain must correspondingly suffer. The mutual relation existing between the different systems, is brought about by the mingling of the nerves, which so completely anastomose with each other that it is impossible to separate them by the nicest dissection. Paradoxical as it may appear, they are separated yet united. The nerves that concern us the most are the lesser splanchnic and solar plexus, both of which belong to the organic nervous system, and are those which control the functions of the kidneys. The nerves that are distributed to the small arteries and capillaries are the *vaso-motor*, and their office is to give "*tone*" to these vessels. They act upon the muscular fibres of the arteries and capillaries, dilating or contracting them according to the stimulus imparted. If we divide the splanchnic nerve, the blood-vessels of the abdominal viscera become dilated, causing congestion. If a stimulus be applied to the same nerve, the same blood-vessels

THE ORGANIC SYSTEM. 41

become contracted. These vaso-motor nerves come from the splanchnic which is a compound nerve, consisting of nerve-fibres from the sympathetic and spinal nervous systems. The vaso-motor nerves control the function of the Malpighian bodies. The nerves from the solar plexus, which is chiefly a sympathetic nerve, are distributed to the uriniferous tubes and control their function. We thus have two sets of nerves distributed to the kidneys, possessing different functions, one to control the elimination of the watery portion of the urine and the other the solid constituents of the same.

THE SYMPATHETIC OR ORGANIC NERVOUS SYSTEM.

Authors have assigned different names to this division of the nervous system. It has been called the "sympathetic nervous system," the "ganglionic nervous system," the "system of organic life," and the "system of vegetable life." We prefer the name "organic," on account of its being the seat of energy that unfolds the organism and sustains it during its physical existence. In its most rudimentary form, it consists of simple nerve granules; in its highest development, of nerve cells, which aggregated together form numerous ganglia. It consists of a series of ganglia united together by connecting nerve branches. There are, also, nerves leading from the ganglia to all parts of the organism; associated with these are nerves from the cerebro-spinal system; the interlacing of these nerves form what are called a plexus of nerves. The nerve fibres arising in the ganglia, according to Beale, surround the ganglionic cells in a spiral form [Fig. 8.]; according to Arnold, they penetrate the cells and are attached to the nuclei, the seat of energy. They evidently enter the cells, which are very minute, being from 1-1200th to 1-2000th of an inch in diameter, in order to receive energy. The origin of this nervous system is differently described by anatomists; some believe it commences in the brain, others in the lower, or coccygial, or sacral region; others believe it arises in the spinal cord. As its function is separate and distinct from the cerebro-spinal system, its source cannot be consistently assigned to brain or cord. Davey believes it to arise from the solar plexus, which he designates as the sun and centre of the sympathetic nervous system; but, as this is merely a plexus of nerves and not a centre of nerve energy, it cannot be the source of a system.

The *semi-lunar* ganglion, which lies immediately below the diaphragm and behind the stomach, is the centre. To describe this system intelligently: The ganglia are arranged in an oval form, one extremity in the brain, the other in the sacral region, and the centre of this oval is the semi-lunar ganglion. Besides this central ganglion there are six single and twenty-four double ganglia, the former being located in the cranium are called cranial ganglia. They consist of the ganglia of Ribes, the ganglia of Laumonier, the lenticular, the phrenopalatine, the otic, and the sub-maxillary ganglia. The twenty-four pairs are called the spinal ganglia, on account of their lying along the spinal column. They are anatomically divided into three servical, twelve thoracic, four lumbar and five sacral ganglia. The *semi-lunar ganglion* is also double, the right being larger than the left ; it is lobulated which gives it the appearance of a series of ganglia aggregated together and presenting a cribiform appearance. From the upper portion of the semi-lunar ganglion arises a number of nerves which unite with nerves from the 6th, 7th, 8th, 9th and 10th dorsal ganglia of the same system, and form the great splanchnic nerve. From the lower portion of the semi-lunar ganglion are given off a number of nerves, which, from their arrangement, have been compared to the rays of the sun. These interlace and form the semi-lunar plexus. Anastomosing with this plexus of nerves are branches of the pneumo-gastric nerve and spinal cord. We will thus perceive that the sympathetic or organic nervous system consists of a series of ganglia, extending from the brain to the sacrum, the centre of which is the semi-lunar ganglion. Each abdominal organ has assigned it a special ganglion from which it is named; the one assigned to the kidneys is the *renal* ganglion, to the liver the *hepatic*, to the spleen the *splenic*, to the heart the *cardiac*, to stomach the *gastric*, etc., etc. To each one is assigned the special function of the organ to which it belongs, the semi-lunar being the co-ordinating centre.

THE ORGANIC SYSTEM.

FUNCTION OF THE ORGANIC NERVOUS SYSTEM.—In man and the higher animals we have two distinct nervous systems; one organic or sympathetic to which is assigned organic life; the other the cerebro-spinal directing animal life. It is the former that concerns us on occount of being the seat of *vital energy* that unfolds the physical organism from the germinal cell and sustains it through its natural life. It is alluded to by some of the older writers, particularly Hippocrates, Aristotle and Plato, as the life principle manifested in organic forms. Aristotle and Plato were the first to speak of it as the basis or groundwork, in studying the functions of the human organism. Serres alluding to the researches of Aristotle and Galen, states, "the method of Aristotle is essentially descriptive, neglecting the function for the form. That of Galen essentially natural, neglecting the form for the function. The first of these methods carried in its train the descriptive sciences. The second led to the general sciences. The truth thus lay in their combination and to Haller we owe the merit of having first discovered this fact. He founded his arguments and opinions upon form and function combined, thus embracing in his method the descriptive as well as general sciences." Among the more modern investigators of the organic life principle stands Harvey, who, quoting Moses as authority, locates "the life of the flesh in the blood, for it is the life of the flesh." The inspired writer was right so far as he expresses it, for the blood possesses vitality which is derived from the organic nervous system. Bichat was the first to bring the subject of vital energy out of chaos. He claimed that the vital phenomena were of two kinds, one corresponding to those functions by which nutrition and growth of individuals are produced, and the other the reproduction of species—both common to all living beings—phenomena which bring into existence living forms and sustain them during physical existence. To Bichat also belongs the credit of originating the idea of the independent existence of the organic nervous system. Since his day, similar views have been advanced by Broussais, Gall, Richerand, Wertzer, and Fletcher of Edinburgh. Previous to the publication of Bichat's treatise on the subject of life and death, it was generally supposed that the entire nervous system was one uniform system. It was from the experiments of Dr. Marshall Hall, that it was fully established that there were three distinct nervous systems, viz.: The brain, spinal cord, and organic or sympathetic. It is somewhat remarkable that Carpenter, Müller, Todd and Bowman have but little to say about the function of the last named system. This reticence has led to an indifference among physicians to the study of the functions of this system in connection with morbid, functional and structural changes in the human organism. It is a well established fact, with many physiologists and physicians of the present day that the organic nervous system controls nutrition by which the human organism is built up and subsequently sustained. It also controls the elimination of disintegrated products from the system.

Is the energy of the organic system vital or physical? It is generally held to be the former. The food used as nourishment when first received is dead matter, the energy that combines the molecules being purely physical. During the process of nutrition it has life imparted to it by the vital energy of the organic nervous system. If this energy be not imparted, it remains crude, lifeless matter, and will be eliminated by the excretory organs, the same as disintegrated matter. Every intelligent physician knows that when the vital forces are depressed, the system cannot be properly nourished. In convalescence from a protracted disease, if food be taken that is not digested and assimilated, the patient cannot be restored to health. Restore the vital energy and administer food sparingly, and the patient will gradually recover. It has baffled the penetration of physiologists and physicians to comprehend the nature of the vital principle. Various names have been given to it. Hunter calls it, "*materia vitæ diffusa;*" Hippocrates, "*impetum faciens;*" Van Helmont, "*archæus;*" Stahl, "*anima.*"

The *vis medicatrix*, or vital principle, is evidently different from the physical energy, being governed by different laws, and manifesting itself differently in the physical organism. To give a detailed view of its character and source, would require more space than is allotted us in this work. We will therefore be compelled to treat it as briefly as possible to be understood. It lies back of the physical, and we must therefore delve into the realms of the unknowable, or unseen universe, where we must study energy and matter in their primitive conditions, and trace them from thence to the external world or visible universe. The more nature is studied the less complex she presents herself to our senses, and we are astonished to find how few and simple are her forces and her laws. Simplicity and concentration of purpose appear to be the peculiar features of the all-wise Being in the unfoldment of creation by His will. Complexity results from the differentiation of forms by the change of matter from an atomic to a molecular condition. Primary matter, which we call *substance* (to distinguish it from physical matter), in its chaotic state, was motionless, unatomic, without form and void, and therefore engulphed in impenetrable darkness. Incorporated with it were two latent energies, one seeking repose in the centre, the other activity at the periphery, called respectively by the ancients, "*attractive power*" and "*repelling motion.*" Before motion, light and life made their appearance, or became manifest, these two were brought into juxtaposition by Deity, creating the law of Harmony, when He ordered light to appear.

By bringing these energies together the *substance* of chaos was broken up into atoms, and, by the antagonism of the forces, motion was produced, and with it light and life. *The atomic substance formed out of the chaotic matter by the antagonism of the twofold force enforced by the law of harmony, and guided by the Divine Will, constituted the beginning of creation.* This substance constitutes the matter of the subjective world out of which the soul is formed; the atomic substance or first matter uniting into molecules constitutes physical matter; this change in substance is caused by the law of harmony yielding or relaxing to what we term equilibrium, lessening the energy of the atoms as well as the light manifested by that energy, and thus changing it from an *imponderable substance* to a *ponderable matter*, and with it the *forces* become *secondary*. This brings us to the objective world with its matter, force,

THE ORGANIC SYSTEM.

and equilibrating law. The forces and matter of the two worlds are distinct, which is entirely owing to the law of harmony becoming modified. The conservation of matter and of force, which is a peculiar feature of this world, cannot be carried into the subjective, the conservation of substance and of energy of the subjective world cannot be carried into the objective; but the subjective forces correlate with the objective forces of *organic* matter.

In order to make this subject clear to the reader: *There was but one original matter which we call substance and from which all matter is derived. There is but one energy in nature, which is twofold, from which all energy is derived; there is but one law in nature from which all law is derived that governs this energy.* Light and heat are not forces, but resultants of force. All objective forces, so called by scientists and physicists, are but one force modified by the law of equilibration. Magnetism, cohesion and chemical affinity, regarded by scientists as three distinct forces, are but one force of a twofold character in equilibrium. Electricity is the same out of equilibrium; light is not visible to the physical eye, it can only be manifested in the simple atomic matter of which the soul is formed. The physical eye only recognizes fire-light, called by the ancients the "*fire-flash.*" If the world is ever destroyed it must be by Deity suspending the law of harmony; if this law should be suspended one moment all cosmical bodies would, in the twinkling of an eye, pass back into chaos. If the reader has been able to follow us, he will perceive that the vital force is purely subjective and unfolds the soul from primary substance, while the physical energy, correlating with the vital, unfolds the physical body. Man thus is a threefold being, a spirit self-encased in the soul and a soul encased in the physical body—and life consists in the correlating of the force of the soul and the organic force of the physical body. Now, we will submit the question: *What is the vital energy of the physical organism ?* It is the energy of the soul correlating with the energy of the nervous system, the former directing the unfoldment of the physical organism in accordance with the type and form of the soul, the physical body being the outward form of the man—the external expression of a spiritual organism. This makes the physical forces of *organic matter* subordinate to the vital, and the energy sustaining physical life a NERVO-VITAL FORCE. At death, the latter withdraws, correlation has ceased, and the physical takes possession of the organic matter and reduces it to its normal chemical constituents. The correlation of the vital and physical forces takes place in the nervous system; so long as harmony exists between them there is normal life-action. When discord exists we have disease which may be of a structural or functional character. The nervous system is the physical medium through which the spirit and soul communicate with the external world. The brain is devoted to psychical, emotional, sensational properties of the spirit, and the organic nervous system to growth, repair and elimination of waste from the organism.

What is disease? It is an impairment or ennervation of the *nervo-vital energies*, operating in and through the organic or sympathetic nervous system, leading to functional derangement or structural changes in the organism. The cause of impairment of physical health is not dependent upon relaxing of the soul's energy, but on the disturbance of the correlation of the nervo-vital forces. The physical organism is merely a dwelling-place for the soul during the spirit's individualization in matter and spiritual probation. The period assigned for man's spiritual probation is three score and ten years; yet how few attain to it, owing to heridicary and acquired causes, impairing the organic nervous system and its energies, and thus preventing the Spirit from sustaining physical life. We shall fully elaborate this subject in a work we are now preparing for the press, entitled, *Life, What it is, Its Source and How it may be Prolonged.*

CHAPTER III.

FUNCTION OR OFFICE OF THE KIDNEYS.

The office of the kidneys is to eliminate some of the morbid products of the blood, the result of disintegration, which is momentarily taking place in the physical organism. The arteries convey the blood to the capillaries to supply the waste that is taking place in the system. The veins receive from the capillaries the blood which contains the product of decay, and convey it to the different eliminating organs, where it is thrown off. The scavengers or purifiers of the blood are the kidneys, lungs, liver, the glands of the alimentary canal and skin. The capillaries are thus the carriers of the blood to the various tissues and organs, to supply the waste that is continually occurring, and they also take up the products of decay and convey them with the blood to the veins. The office of the kidneys is to eliminate the azotized or nitrogenized products; the liver the carbo-hydrogen, and the lungs carbonic acid. If these decayed elements should be retained in the blood a short time, they poison the nervous centres, and cause death. It is therefore of the utmost importance that the eliminating organs should be in a healthy and active condition, to cleanse the blood of these morbid products. The great secret of health is to take into the system sufficient food, and no more, to compensate for the waste, and to keep the excretory organs in a healthy condition, to remove the impurities from the blood. If we could maintain this balance (in a normal or healthy constitution) between repair and waste, and not overtax our vital energies, there would be no difficulty in attaining to the age alotted to man, if not going considerably beyond it. In order to understand the functions of the kidneys, it will be necessary for us to familiarize ourselves with the structure, which we have fully described in a previous chapter. It is in the Malpighian tufts and uriniferous tubes of the cortical portion of the kidney where this function is performed. Dr. Bowman in an essay, published in 1842, was the first to advance a true theory of the function of this portion of the kidney.

On account of the clearness with which he has presented this subject, we think it advisable to quote him:—

"Reflecting on this remarkable structure of the Malpighian bodies, and their singular connection with the tubes, I was led to speculate on their use. It occurred to me that as the tubes and their plexus of capillaries were probably, for reasons presently to be stated, the parts concerned in the secretion of that part of the urine to which its characteristic properties are due (the urea, lithic acid, etc.), the Malpighian bodies might be an apparatus destined to separate from the blood the watery portion. This view, on further consideration, appears so consonant with facts, and with analogy, that I shall in a few words state the reasons that have induced me to adopt it. I am not unaware how obscure are the regions of hypothesis in physiology, and shall be most ready to renounce my opinion if it should be shown to be inconsistent with truth.

"In extent of surface, external structure, and the nature of its vascular net-work, the membrane of the uriniferous tubes corresponds with that forming the secreting surface of other glands. Hence it seems certain that this membrane is the part specially concerned in eliminating from the blood the peculiar principles found in the urine. To establish this analogy, and the conclusions deduced from it, a few words will suffice: 1. The extent of surface obtained by the involution of this membrane will by most be regarded as its self-sufficient proof. 2. Its internal surface is conclusive. Since epithelium has been found by Purkinje and Henle in such enormous quantities on the secreting surface of all true glands, its use cannot be considered doubtful. It never forms less than 19-20ths of the thickness of secreting membrane, and in the liver it even seems to compose it entirely, for there I have searched in vain for a basement tissue like that which supports the epithelium in other glands. The epithelium, thus chiefly forming the substance of secreting membrane, differs in its general characters from other forms of this structure. Its nucleated particles are more bulky, and appear from their refractive properties to contain more substance, their internal tissue being very finely mottled when seen by transmitted light. In these particles the epithelium of the kidney tubes is eminently allied to the best marked examples of glandular epithelium. 3. The capillary net-work surrounding the uriniferous tubes is the counterpart of that investing the tubes of the testes, allowance being made for the difference in the capacity of these canals in the two glands. It corresponds with that of all true glands in lying on the deep surface of the secreting membrane, and its numerous vessels everywhere anastomosing freely with one another.

"These several points of identity may seem too obvious to be dwelt upon, but I have detailed them in order to show that, in all these respects, the Malpighian bodies differ from the secreting parts of true glands. 1. The Malpighian bodies comprise but a small part of the inner surface of the kidney, there being but one to each tortuous tube. 2. The epithelium immediately changes its character as the tube expands to embrace the tuft of vessels. From being opaque and minutely mottled, it becomes transparent, and assumes a definite outline; from being bald it becomes covered with cilia (at least in reptiles, and probably in all classes); and in many cases it appears to cease entirely a short way within the neck of the Malpighian corpuscle. 3. The blood-vessels instead of being on the deep surface of the membrane, pass through it, and form a tuft on its free surface. Instead of the free anastomoses elsewhere observed, neighboring tufts never communicate, and even the branchlets of the same tuft remain quite isolated from one another.

"Thus the Malpighian bodies are as unlike, as the tubes passing from them are like, the membrane which, in other glands, secern its several characteristic products from the blood. To these bodies, therefore, some other and distinct function is with the highest probability to be attributed.

"When the Malpighian bodies were considered merely as convoluted vessels, without any connection with the uriniferous tubes, no other office could be assigned them than that of delaying the blood in its course to the capillaries of the tubes, and the object of this it was impossible to ascertain. Now, however, that it is proved that each one is situated at the remotest extremity of the tube, that the tufts of the vessels are a distinct system of capillaries inserted into the interior of the tube, surrounded by a capsule formed by its membrane, and closed everywhere except at the orifice of the tube, it is evident that conjectures on their use may be formed with greater plausibility.

"The peculiar arrangement of the vessels in the Malpighian tufts is clearly designed to produce a retardation in the flow of blood through them, and the insertion of the tuft into the extremity of the tube is a plain indication that this delay is subservient in a direct manner to some part of the secretive process.

"It now becomes interesting to inquire, in what respect the secretion of the kidney differs from that of all other glands, that so anomalous an apparatus should be appended to its secreting tubes. The difference seems obviously to lie in the quantity of aqueous particles contained in it, for how peculiar soever to the kidney the proximate principles of the urine may be, they are not more so than those of other glands to the organs which furnish them.

"This abundance of water is apparently intended to serve chiefly as a menstruum for the proximate principles and salts which this secretion contains, and which, speaking generally, are far less soluble than those of any other animal product. This is so true, that it is common for healthy urine to deposit some part of its dissolved contents on cooling.

"If this view of the share taken by the water be correct, we must suppose that fluid to be separated either at one point of the secreting surface along the proximate principles, as has hitherto been imagined, or else in such a situation that it may at once freely irrigate the whole extent of the secerning membrane. Analogy lends no countenance to the former supposition; while to the latter, the singular position and all the details of the structure of the Malpighian bodies, gives strong credibility.

"It would, indeed, be difficult to conceive a disposition of parts more calculated to favor the escape of water from the blood than that of the Malpighian body. A large artery breaks up, in a very direct manner, into a number of minute branches, each of which suddenly opens into an assemblage of vessels of far greater aggregate capacity than itself, and from which there is but one narrow exit. Hence must arise a very abrupt retardation in the velocity of the current of blood. The vessels in which this delay occurs are uncovered by any structure. They lie bare in a cell from which there is but one outlet, the orifice of the tube. This orifice is encircled by cilia in active motion, directing the current to the tube. These exquisite organs must not only serve to carry forward the fluid already in the cell, and in which the vascular tube is bathed, but must tend to remove pressure from the free surface of the vessels, and so to encourage the escape of their more fluid contents. Why is so wonderful an apparatus placed at the extremity of each uriniferous tube, if not to furnish water to aid in the suppuration and solution of the urinous products from the epithelium of the tube?"

The theory advanced by Dr. Bowman that the uriniferous tubes secrete the solid constituents of the urine, is verified by examining the urine of the *boa*, which is almost solid. The solidity of the urine of the *boa* is owing to the artery supplying the kidneys being very small compared with other animals, and eliminating but a very small quantity of water, barely sufficient to carry off the solid material secreted by the tubes. Observations made by Dr. Johnson are also confirmatory of Dr. Bowman's theory. He states: " In examining the kidneys of a person who died with jaundice, and in whose urine there had been a large quantity of bile, I observed that the tubes were stained of the deep yellow color of bile in their epithelial cells, and that the yellow color ceased absolutely at the neck of the Malpighian corpuscle, and in no one instance did it affect any part of the tissues of the Malpighian bodies." We may therefore conclude, without further testimony from the observations of Dr. Bowman and the confirmation of modern physiologists, that the solid constituents of the urine are secreted by the uriniferous tubes of the kidneys. The straight portion of the tubes of the pyramid have no secreting power assigned them, although they are lined by the same character of epithelium as the cortical portion. They are merely excretory ducts for the purpose of conveying off the excretory products of the tortuous portion of the tubes and the water eliminated by the Malpighian bodies into the pelvis of the kidneys.

The kidneys perform another important office in the human economy, besides eliminating nitrogenized products. They act as a safety-valve in regulating the quantity of water in the blood. They, with the lungs and skin, are the only means through which the system has the power of removing the water from the blood. The amount of fluid that passes off in the form of insensible perspiration and by the lungs, according to Seguin, is eighteen grains per minute, eleven grains of which is thrown off by the skin and seven grains by the lungs. The maximum quantity eliminated by the skin and lungs in twenty-four hours is five pounds, and the minimum quantity, one pound and two-thirds. The exhalation of the skin varies according to the condition of the atmosphere; when it is moist, it is diminished, when dry, it is increased. When the temperature is high and the atmosphere dry, a large amount of moisture is thrown off in the form of sensible perspiration, and with it there escapes a large amount of free caloric relieving the system from oppression which might otherwise occur. We are all aware of the relief free perspiration affords in extremely warm weather, and how soon a simple fever may be broken up by inducing a copious perspiration. For many years, I have been in the habit of breaking up catarrhal and remittent fevers by bringing about a free perspiration and keeping it up until the fever subsides, which it will usually do in from six to twenty-four hours. In this plan of treatment the head should always be kept cool by the application of ice.

Several years ago, when the relapsing fever prevailed in Philadelphia, the *crisis* was brought about on the seventh day by Nature causing the patient to perspire copiously, and the kidneys to act freely. If the first attack, the patient would be apparently well and out in from twenty-four to forty-eight hours, to be again stricken on the seventh day, with the symtoms aggravated, and when repeated the third time, it was generally fatal. After losing one case during the third attack, I anticipated Nature by producing at the commencement of the disease a copious perspiration, and keeping it up for forty-eight hours. I never lost a case afterwards, neither did I have any cases to relapse. It is by perspiration of the skin that we get rid of the free caloric, which, if retained, brings about congestion of some one of the internal organs. What is commonly known as *coupe de soleil*, or *sun-stroke*, is the overheating of the blood, producing cerebral congestion. We thus perceive the danger of suddenly checking perspiration, which is frequently and ignorantly done by sitting in a draught, or by removing superfluous clothing, when overheated. The blood-vessels of the skin become contracted by sudden change of temperature, forcing as uperabundance of blood upon the kidneys or lungs, etc., causing congestion. Most persons are aware that less water is passed by the kidneys in summer than in winter. This is owing to the blood-vessels becoming contracted, and perspiration lessened, causing the blood to be directed toward the kidneys, producing a more rapid discharge of water from the Malpighian bodies. Persons affected with Bright's Disease should be exceedingly careful to protect the surface of the body in summer as well as in winter, for fear of increasing congestion of the urinary organs, which always attends Bright's Disease. Cutaneous perspiration has no influence on the amount of solid matter secreted by the kidneys; it depends upon the waste taking place in the system, and the excess of food taken to supply the wants of nutrition. When there is too much to compensate for the waste, the excess is not assimilated, but undergoes rapid retrograde metamorphosis, and must be eliminated by some one of the excretory organs. It is by excess of feeding, and the drinking of spirituous liquors, that the gormandizer and inebriate overtax the liver and kidneys, producing either organic or functional disease, and leading to much suffering, if not to premature death.

Under ordinary circumstances, each excretory organ is limited to its own special function, yet there are certain *complementary* relations between them which enable them to render assistance to each other. For example, when respiration is active, the lungs eliminate more carbonic acid, and the liver secretes less bile, it consisting largely of carbon. If respiration be diminished, there is less carbon thrown off by the lungs, which is eliminated with hydrogen, in the form of bile. We find similar *complementary* relations existing between the skin and the kidneys, to which we have already alluded. While speaking of the mutual relations existing between the different organs, it will not be out of place to allude to what are called *vicarious* excretions. It is the metastasis or removal of an excretion from its natural outlet through some organ possessing entirely different function. For example, urine has been known to have been eliminated from the bowels, eyes, breast (of the female), stomach, mouth, nose, ears, skin, etc. Dr. Lacock has collected a large number of such cases, which we give in the following table:—

Vomit	Stool.	Ears.	Eyes.	Saliva	Nose.	Breast	Navel	Skin.	Total.
33	20	4	4	5	3	4	34	17	124

A familiar form of excretory metastasis is

found in jaundice, where the bile is secreted by the liver, absorbed in the blood, and eliminated by the skin, kidneys, etc.

SECRETION OF URINE.—To understand the morbid condition of the urinary organs, we must be familiar with their normal functions. Fresh, healthy urine is a clear, amber yellow liquid, and of a peculiar, disagreeable odor, with a well marked acid reaction. If permitted to stand some time a slight cloudiness appears, consisting of mucus, which gradually settles at the bottom of the vessel. Very soon a different, more unpleasant odor is developed, when the urine becomes alkaline, which is owing to the decomposition of urea into carbonate of ammonia, and a precipitate of earthy phosphates.

Healthy urine may become turbid soon after being passed, on cooling, which is due to the precipitation of urates of soda and ammonia, which are very soluble in warm water, but insoluble in cold water. It is only when they are slightly in excess that they are precipitated. *If the urine should be turbid on passing it is considered abnormal.* The quantity of urine passed in twenty-four hours varies, depending upon the amount of liquid taken into the system, and the external temperature. If the surface of the body should be chilled, and perspiration become partially checked, there is more water passed by the kidneys. According to the observations of Dr. Prout, about 30 ounces is the amount passed in summer, and 40 ounces in winter. The *specific gravity* of urine varies with the amount of solid food taken, and the amount of waste going on in the organism. The average, taking the whole year, according to the same writer, is about 1020, rising in summer to 1025, and falling in winter to 1015. This variation, as we have previously stated, is owing to the much larger amount of fluid eliminated by the skin during the warm summer months in the form of perspiration than during the cold, chilly months of fall, winter and spring.

The amount of solid matter in 1000 parts of urine fluctuates widely, the difference being due to diet, waste taking place in the system, and the amount of water eliminated.

The following tables have been constructed by Simon, from Leeman's Analysis:

Urine.	Mixed Diet.	Animal Diet.	Vegetable Diet.
	Grammes*	Grammes.	Grammes.
Amount in 24 Hours	1057.8	1202.5	909.
Specific Gravity	1022.	1027.1	1027.5
Solid Residuum from 1000 Parts of Urine Passed	65.82	75.48	66.41
Solid Matter Passed in 24 Hours	67.82	87.44	59.23

Urea.	In 100 parts of solid residuum.	Daily amount in grammes.
During a mixed diet,	46.230	32.498
During an animal diet,	61.297	53.198
During a vegetable diet,	39.086	22.481

Uric Acid.	In 100 parts of solid residuum.	Daily amount in grammes.
During a mixed diet,	1.710	1.183
During an animal diet,	1.674	1.478
During a vegetable diet,	1.737	1.021

It will be observed from these tables that the amount of *urea* is always diminished by vegetable diet, while the quantity of uric acid is not materially affected. A knowledge of this fact is of the utmost importance in the treatment of Bright's Disease. Vegetable diet considerably increases the amount of extractive matter, which is diminished in animal diet. The amount of

---

*A gramme is equal to 15.4 grains Troy.

phosphates, lactic acid and lactates are but slightly changed by an animal or mixed diet.

Constituents.	Mixed Diet.	Animal Diet.	Vegetable Diet.	Non-nitrogenized Diet.
	Grammes	Grammes	Grammes	Grammes
Solid constituents, . .	67.82	87.44	59.24	44.68
Urea, ...	32.50	53 20	22.48	15.41
Uric Acid, ...	1.18	1.48	1.02	0.73
Lactic Acid and Lactates,	2.72	2.17	2.68	5.82
Extractive matters, . .	10.49	5.20	16.50	11.85

We observe that urine consists of a liquid and solids, which are produced by two distinct structures, but closely united. They comprise the Malpighian bodies and uriniferous tubes, the functions of which we will speak of separately.

1. MALPIGHIAN BODIES.—The Malpighian tufts or glomerules consist of capillaries twisted upon themselves, forming a round mass or ball, and enclosed within the upper and dilated portions of the uriniferous tubes [Figs. 3 and 4]. These capillary vessels are minute tubes lying between the arteries and veins, and so numerous that it would be impossible to insert the point of a needle in

any part of the soft tissues without injuring them. It will not be amiss here to state, briefly, the relation existing between the capillaries, arteries and veins. The arteries arise from the ventricles of the heart in two branches, one going to the lungs, and called the pulmonary artery, the other to the general system, called the aorta. They divide and subdivide, and finally terminate in capillaries. The veins arise from the capillaries and convey the blood back to the heart; the capillaries are, therefore, a connecting link between the arteries and veins. All of these vessels have elastic walls, consisting of muscular fibres and muscular cells, which are well defined in the large arteries, but gradually diminish as they emerge into the capillaries, whose walls are exceedingly thin and permeable, permitting the fluid portion of the blood to pass through them into the surrounding tissues or " lymph space." *There is a normal constriction in the capillary walls which limits the amount of fluid passing through them.* The area of the aorta is much less than the area of its branches. Vierodt has estimated the area of the capillaries, and makes them several hundred times larger than the area of the aorta. They may be compared to an inverted funnel, with its apex at the aorta. To the arteries and capillaries are distributed nerve fibres, called *vaso-motor* nerves. These nerves control the calibre of the capillaries, as we have shown in a previous chapter. If there is a deficiency of nerve-energy, the walls are relaxed, circulation sluggish, and the parts assume an increased redness. The emotions frequently bring about this condition, producing what is called "blushing." The opposite condition takes place in what is called "palor." If the splanchnic nerve be divided, the walls of the capillaries suddenly relax, followed by a copious filtering of water from the blood, as in hydrura or poly-urea. If a stimulus be applied to the divided nerve, the relaxation of the tufts is overcome and the copious flow ceases. An injury to the cerebellum will give rise to an increased flow of urine; a puncture of the fourth ventricle of the brain or a mechanical injury to the fourth thoracic ganglion causes a large quantity of sugar to appear in the urine, with a rapid increase of fluid. This is known as artificial diabetes. Injuries to the pneumo-gastric or par-vagum nerve, or to the medulla oblongata cause a rapid flow of urine. These injuries impair the function of the vaso-motor nerves, which are branches of the splanchnic nerve, containing motor fibres from the spinal cord. From what has been stated, it can readily be perceived that the vaso-motor nerves are spinal branches associated with sympathetic nerve branches, controlling capillary circulation, and, therefore, modifying the quantity and character of the water eliminated by the Malpighian tufts, which is not a secretion, but simply a filtration through the capillary walls. We have, then, two sets of nerves distributed to the kidneys: one vaso-motor, which are inhibitory and come from the spinal system, and yet are associated with the splanchnic nerve, a branch of the sympathetic, and nerves from the solar plexus which control the secretion of the uriniferous tubes. It is believed, and so stated, by physiologists, that filtration depends upon arterial pressure. It may, to a limited extent. The arrangements of the capillaries in forming the glomerules is to economize space and give more filtering surface. In the *boa*, the renal artery is very small, and the glomerules correspond in size with the artery. The vaso-motor nerves give tone to the walls of the capillaries, which a moderate irritation increases. If pushed too far, the nerve becomes exhausted and relaxation ensues, the tone is gone, and the filtration of water is increased. I am, therefore, of the opinion that arterial pressure does not facilitate filtration in the glomerules, but relaxation of the capillary walls, by a diminution of vaso-motor energy, causes it.

2. *Uriniferous Tubes.*—We have shown that the office of the Malpighian bodies is to eliminate the superabundance of water in the blood which is accomplished by a process of filtration through the walls of the capillary tufts; the quantity eliminated depending upon the tonicity of the walls, imparted by the vaso-motor energy.

The function of the uriniferous tubes is very different from that of the Malpighian tufts, being performed by an epithelium structure resting on the basement membrane and lining the tubes. The office is one of excretion, and is a nervo-vital process, the energy being imparted through the solar plexus. In the kidneys there are *two perfectly distinct systems of capillary vessels.* The first consists of the Malpighian tufts which are immediately connected with the arteries. The second envelopes the uriniferous tubes, forming a plexus around the tubes and so arranged as to form a basis for the basement membrance of the tubes; the capillaries then terminate in veins [Fig. 3]. We have already given a description of this membrane in a previous chapter, and all that is necessary to say here, is that the material of which it is composed is an elimination from the capillaries, which, undergoing a vital change from energy imparted by the ganglionic nervous system, causes the exuded fluid to become granular. These granular particles unite to form the nuclei, and these possess the inherent vital energy to construct the epithelium cells that rest upon the basement membrane. Nerve fibres have been traced into this membrane, and are supposed to extend up into the epithelium cells and to be attached to the nuclei. This is not requisite; the nerve fibres convey an energy to the blood lymph that is eliminated from the capillaries, and in revitalizing it produce the granular particles which aggregate and form muclei, and the muclei possess the power to create cells. Therefore, all that is necessary is for the nerves to enter the fluid mass and create vitalizing centres, which are the granules. All secretory and excretory cells are frequently removed, for as soon as the cell contents are changed into the solid constituents of the urine the cell disintegrates and another immediately takes its place from the basement membrane. In all secreting and excreting organs, the formation and disintegration of cells depend upon the activity of the functions of the organ. In the stomach, during digestion and in the villi of the intestines during the absorption of chyle, the functions are exceedingly active, and the formation and disintegration of cells is very rapid. The change of cells in the kidney tubes depends upon the amount of serum eliminated from the vessels, and the amount of nerve energy imparted to it. For example, in fevers and inflammatory condition of the kidneys both are increased, and a greater amount of solid constituents is excreted and therefore a more rapid formation and disintegration of cells must take place. If this vital nerve energy should be diminished, and less blood serum be eliminated, the specific gravity of the urine becomes lessened, owing to a diminution of the solid constituents. We may regard all epithelium cells in any part of the system as small laboratories and the work-master or chemist the nucleus. If the master be sick or out of sorts (and this will depend upon the condition of the nervous system that gives him his energy) the work is imperfectly performed and the result, disease. In proof that the nervous system is the source of energy, by stimulating the nerve branches of a gland the secretion is rapidly increased and is much richer in constituents. If the nerve fibres be divided the secretion stops entirely, or becomes very fluid. We shall show when we speak of the cause of Bright's Disease, that in the organic nervous system lies the cause of this much dreaded and inveterate malady. If a solution of indigo-carmine be injected into the veins of the kidneys of an animal in which the urinary flow has been arrested by an incision of the medulla oblongata, the carmine will be found to have passed through the walls of the capillaries, and been diffused through the basement membrane, taken up by the cells and eliminated into the tubes. No trace of it, however, can be found in the Malpighian tufts. This shows the method nature adopts for the elimination of morbid products from the blood which is accomplished through cell structure, resting upon and formed from a basement membrane. The uriniferous tubes are secretory glands, taking from the blood lymph, eliminated by the capillaries, the waste products and by nervo-vital chemistry changing them into urea, uric acid, extractive matter, etc.

How are the disintegrated products of the tissues, which are so varied, distributed by the blood-vessels to the organs assigned for their elimination? For example, the kidneys take from the blood the nitrogenous products of waste, the liver the carbo-hydrogen products of the same. This question has yet to be answered by physiologists. It is well known to the physician, that if tartar-emetic or ipecac be injected into the bowels or into the blood, it will produce nausea and vomiting as if taken into the stomach. If urea be introduced into the stomach or injected into the circulation, it makes its appearance in the kidneys for elimination; if Cantharides, turpentine, and the various balsams be taken into the system, they are likewise eliminated by the kidneys. The nervo-vital energy possesses the power of establishing affinities or sympathies in the excretory organs by which they attract certain substances from the blood that, if retained, would be injurious to the system. It is upon this basis that homœopathists prove their remedies. For example, in desiring to know the properties of a remedy, they give it to several persons in health unknown to each other; when symptoms are produced alike in all these provers, it shows that the remedy has a special affinity for certain parts of the system, and the symptoms that are produced in these parts are their guide in selecting the remedy when a diseased condition exists presenting similar symptoms. Certain organs evince affinities for certain remedies, and these affinities are produced by the vital energy of the soul, correlating with the organic nerve-energy.

CHAPTER IV.

ABNORMAL CONDITION OF THE KIDNEYS IN BRIGHT'S DISEASE.

The structural changes that take place in Bright's Disease during its progress have been classified under different heads. Reinhardt and Frerichs believe the disease results from inflammation of the kidneys, and have divided it into three stages; 1. The inflammatory stage; 2. The exudation, in which is observed a fatty degeneration of the epithelium lining of the tubes and the glomerules; 3. Abnormal growth of the connecting tissues, terminating in atrophy. The German and French pathologists insist upon the unity of the disease; the English pathologists, on the contrary, deny the unity, maintaining that the disease comprises several distinct *forms*. They base their opinion on the cause, the symptoms, and the pathological changes that occur. Dr. Johnson, in his work on disease of the kidneys, published in 1852, and his lectures on the same subject, published in

1873, divides this disease into acute and chronic forms. The former represents the inflammatory stage, and the latter the pathological changes that occur during the progress of the disease. The first change, involving the epithelium lining of the uriniferous tubes and the glomerules, constitutes the red granular kidney. When the parenchymatous structure (the body of the kidney) becomes involved, we have the second stage of the disease; this constitutes the large, white kidney. When the organized tissues become absorbed, and atrophy takes place, it comprises the third stage, and is called the lardaceous or waxy kidney. Roberts, Dickson, Stewart, and others have somewhat modified the division of Johnson, without materially differing with him, or throwing light upon the subject. We therefore deem it unnecessary further to refer to these writers, as it will only confuse the reader, instead of enlightening him. There is no doubt but that Bright's Disease assumes two distinct forms, acute and chronic, as is the case with a large majority of diseases, providing the former is allowed to pursue its course. In the acute form there is a higher grade of inflammation, which is due to an active impairment of the nervo-vital energy of the vaso-motor nerves, causing a relaxation of capillary blood-vessels of the kidneys, and consequently an accumulation or superabundance of blood in the parts involved. Nature endeavors to overcome the nervous shock, and if she succeeds the acute stage subsides. If she only partially succeeds, inflammation becomes somewhat lessened, and the disease gradually passes into what is called the chronic form. We cannot conceive why there should be a conflict of opinion between the English and Continental pathologists, for Bright's Disease, when it becomes manifested in the kidneys, and is allowed to continue, causes various pathological changes. In the acute form it is more functional than organic, but if permitted to continue, the changes described by the English physicians occur. Yet it is one and the same disease. Although Bright's Disease is always accompanied by inflammation, the mere fact of inflammation does not constitute Bright's Disease. Inflammation is an ennervation of the vaso-motor nerves, permitting a relaxation of the capillaries, and causing an increased flow of blood to the parts, consequently an increased redness and oxidation, therefore an increase of temperature. All this may occur without the elimination of albumen, which is the characteristic feature of Bright's Disease. To produce this inveterate complaint something more than inflammation must exist. *In Bright's Disease we have not only an ennervation of the vaso-motor nerves* (the cause of inflammation), *but an ennervation of the nerves of the solar plexus, causing a change in the secretion of the cellular structure of the uriniferous tubes, and a change in the nutrition of the entire kidney, and this constitutes Bright's Disease.* We therefore are of the opinion that Bright's Disease assumes two distinct forms, acute and chronic, which we shall treat of separately.

1. THE ACUTE.—This form is not near so common as the chronic; it occasionally follows the acute stages of scarlet fever, small-pox and diphtheria. In the first instance, it sets in about the end of the second week, and in cases where the eruption is not well defined, and is aggravated

PATHOLOGICAL CHANGES.

by exposure to cold and dampness. In small-pox, it occurs during the disquamation period, and in cases where the system has received a severe shock from the poison. In diphtheria, it makes its appearance in cases where the system is very much reduced. The acute form occasionally occurs during pregnancy, but generally terminates with it. The vegetable and mineral poisons, and the too free use of alcoholic drinks, bring about a similar condition of the kidneys. The pathological changes observed in the acute form are principally confined to the cortical portion of the kidneys. The Malpighian bodies present a dark-red appearance; their epithelium covering is more or less detached, and the capillary walls thickened. The epithelium lining of the tubes is more prominent and elevated, giving evidence of overgrowth. The tubes become choked with exuded epithelium cells mixed with more or less blood. The cells lining the tubes may be natural, but opaque, and disintegrate rapidly and are easily detached from the basement membrane. There is an exudation in the form of transparent cylinders of various sizes in the tubes that have shed their epithelium. These abnormal changes seriously impair the function of the kidneys, and, if not arrested, death ensues from blood-poisoning. In the acute form, the disease may not be so active as to produce death; the inflammation gradually subsides, and merges into the chronic form, or the disease may be arrested, the inflammation entirely removed, the nervous energy re-established and the patient restored to health.

2. THE CHRONIC.—This form may result from the acute, or the disease may assume the chronic form from the start, and develop so gradually and imperceptibly as to be unnoticed by the patient, his friends or his physician. Most writers describe three stages of this form of Bright's Disease based upon pathological changes observed after death :

1. That of inflammation with commencing exudation, frequently called *tubal nephritis* on account of the uriniferous tubes being the principal seat of diseased action.

2. *Granular Degeneration*, in which exudation has taken place, which subsequently extends to the intertubular matrix.

3. *The Waxy or Lardaceous Form.*—In this stage of the disease, the kidneys undergo degeneration and atrophy.

These three forms or types of disease may be summed up as follows: 1. The smooth white kidney. 2. The granular mottled kidney. 3. The waxy or amyloid kidney.

1. THE INFLAMMATORY OR FIRST FORM OF CHRONIC BRIGHT'S DISEASE.—The morbid changes are very similar to those observed in the acute form; the inflammation is less active. If we examine the cortical substance with a magnifying glass, we find scattered through it hæmorrhagic spots, the result of congestion or engorgement of the glomerules, giving it a mottled or granular appearance. We also find an increase in the number and size of the oil-globules of the epithelium lining of the tubules. As the disease advances the tubes become distended and filled with cast-off epithelium oil-globules and blood corpuscles, all matted together, completely blocking a portion of the tubes, while others are comparatively free, their epithelium presenting a healthy appearance. In the more advanced stage

of the disease the epithelium lining of the cortical portion of the tubes is destroyed, and fibrous casts of these tubes are thrown off. The medullary portion and the glomerules are not seriously involved.

The hyperplastic process takes place in the basement membrane, and not in the epithelium cells as is generally supposed, and is dependent upon *ennervation of nervo-vital energy*, causing a rapid development of nerve cells and nuclei and an increase of oil-globules. Here growth over-reaches the boundary of harmony, by which it is guided in health. If this condition continues the epithelium is imperfectly formed, the walls of the tubes lined with the granular mass instead of cells. Fortunately for the patient this condition does not exist in all the tubules. If it did the excretory function of the kidney would be suspended and the nitrogenized products of waste retained in the blood, and the patient die of uræmia or blood poison. Occasionally in this stage of the disease, the structural change is not confined to the tubules, but extends to the fibrous structure in which fatty degeneration is found, causing an enlargement of the organs to double their natural size, giving the kidneys a yellowish-white appearance—hence the name: *the smooth white kidney.*

This form of Bright's Disease is generally traced to some exciting cause, as cold, consumption, pregnancy, scarlet fever, etc. In one hundred and six cases examined by Dickson, the average age was about twenty-eight years.

The urine is scanty, pale and cloudy, specific gravity normal; if permitted to stand, it deposits a sediment consisting of tube-casts, epithelial, fatty matter, granular and cretaceous substance. Dropsy is almost always attendant on this form; the skin is white, flesh soft and flabby, face pale and puffy. There is seldom heart trouble accompanying this type of the disease, but it is a frequent attendant in the granular form. There is very frequently, however, secondary inflammation as pneumonia, peritonitis, etc. It runs its course, however, in a much shorter time than the granular form, usually in from six to nine months. Occasionally, however, it may be prolonged for several years.

2. THE GRANULAR OR SECOND FORM OF

CHRONIC BRIGHT'S DISEASE.—The kidneys are diminished in size and weight. The surface is very unlike that in the form just described, being uneven or rough, owing to the number of small elevations, varying in size from a pin's head to a pea. The capsule adheres closely to the adjacent surface, and presents a granular appearance. The texture of the kidney is tough and unyielding. If the cortical portion be examined, it will be found very much contracted, forming a thin rim around the pyramids, not more than one-sixth of an inch in thickness. If we examine it with a microscope, we find it has undergone a marked morbid change. The Malpighian corpuscles are diminished to half their size, and closely crowded together by a fibrous and granular investment. Portions of the tubes are denuded of their epithelium, and so contracted as to present mere tubular threads. Others are blocked up with broken-down epithelium and oil-granules. What is remarkable, normal or healthy tubes are found side by side with those that have become almost obliterated. It is the generally accepted opinion that the morbid process commences in the epithelium cells lining the tubes. Dickson and Stewart maintain that it occurs in the inter-tubular space, which they consider an overgrowth or hypertrophy of the fibrous tissues which compresses the Malpighian corpuscles. In this way, they account for the contraction of the cortical portion of the kidney. As the disease progresses, the pyramids become involved, cysts are frequently met with, which vary in size from a pin's head to a pea. They are evidently formed by the obstruction and distention of the uriniferous tubes, for they are found lined with an epithelium and filled with an exudation that is not urinous, but albuminous. These cysts are generally found in the cortical substance, and occasionally in the cones.

This type of the disease makes its appearance insidiously, without any definite or exciting cause. It advances slowly, and may exist for months, and even years, and the sufferer be unconscious of it until it is discovered by his physician. The urine is generally at first copious and of low specific gravity. As the disease advances the quantity gradually diminishes, the albuminous deposit is slight and occasionally absent. In the majority of cases there are to be found in the urine epithelium and granular casts, which are eliminated from the uriniferous tubes. In many, perhaps half of the cases, there is no dropsical effusion; when it is present, it is generally limited to the ankles and under the eyes. Such persons are generally pale and anæmic, with countenance pinched, and the skin presents a sallow appearance. Hypertrophy, or enlargement of the heart, is a very frequent accompaniment, and from half to two-thirds of the cases are afflicted with this form of cardiac difficulty. There is a general cachectic condition of system, showing conclusively an impaired nutrition. Out of two hundred and fifty cases examined by Dickson, the average age was about fifty years.

3. THIRD FORM OF CHRONIC BRIGHT'S DISEASE.—WAXY OR LARDACEOUS KIDNEY.—In this form of the disease the kidney is usually enlarged. The cortical portion presents a whitish or light yellow color. If a section be examined with a microscope, it has a waxy appearance, and is confined almost exclusively to the muscular coats of the blood vessels. The Malpighian cor-

puscles are the first to become affected with this waxy deposit. As the disease advances, the arteries and capillaries of the entire kidney become involved; the epithelium of the tubules is contracted and infiltrated with fat corpuscles; the liver and spleen are usually found to contain the same fatty deposit. The true nature and source of the waxy material has not been definitely determined. Dickson supposed it to be a species of fibrin, containing less alkali, and a larger proportion of earthy salts. Virchow believed that it belonged to the same class of substances as salt and cellulose, on account of its yielding a violet color when combined with sulphuric acid and iodine. It is evidently of a protean character, for it contains the same amount of nitrogen as the protean compounds. Like the granular form of this disease, it makes its appearance in cachectic persons, and comes on insidiously. In the commencement the urine is generally copious, and as it advances it diminishes in quantity, while its specific gravity increases. The albumen at first is but slight, but becomes quite profuse as the disease advances. The urine is pale, and contains epithelium cells and fat corpuscles. This form of the disease may be diagnosed by the abundance of albumen in the urine, the profuse dropsical effusions, and the general wasting of the system. It is not unfrequently complicated with syphilis or phthisis, and, when the system has been reduced, by constant suppuration.

CHAPTER V.

What is Bright's Disease? Its Curability.

Bright's Disease is generally understood to be an inflammation of the kidneys, either acute or chronic, and characterized by albuminous deposits in the urine. We do not deny the correctness of this so far as the kidneys are concerned, but it declares only the effect of a cause which must be sought for, ascertained and controlled before we can expect to place it amongst the curable diseases. Therefore to designate it as inflammation of the kidneys with albuminous deposits in the urine, does not give a correct idea of its character, magnitude or cause.

The main object of this work is to call the attention of those suffering with this troublesome complaint to some new facts as to the cause, and afford them hope that, if the disease be treated in time, it will *not* be found incurable, as it is supposed to be by physicians generally. It is almost a universal opinion, in the profession, that the disease, in its chronic form, is incurable, and they are often too free and decided in giving an unfavorable prognosis as soon as they ascertain that a person has Bright's Disease. This belief also prevails with the community, and when confirmed by the physician the patient gives up all hope of recovery, which naturally depresses the vital energies, and thus hastens the progress of the disease. On the contrary, we do not regard it as right and proper for the physician to buoy up the patient with false hopes. We have known instances in the last two years where persons afflicted with Bright's Disease could have been cured, or, if not cured, life prolonged, where the attending physician insisted (when the patient contemplated change of treatment) that there was not the slightest hope for a cure. They asserting that the best authorities in the profession had pronounced the disease incurable—that settled it. Two of the cases to which we have reference, are in their graves, while those who were quite as severely afflicted, if not more so, are enjoying comparatively good health.

THE VITAL FORCE.

Disease being an abnormal life action, it will be necessary, in order to understand it properly, to have a knowledge of normal life, the former being merely a deviation from the latter. As life cannot be manifested without motion there must be, in the universe, a life energy. Many advanced scientific thinkers maintain that life results from molecular change.

This cannot be correct, for the organized principle or vitalized energy *controls* the physical force. We find when there is an expenditure of nerve-force, disintegration of the tissues ensues, corresponding to the amount of vital energy manifested. If the brain be overactive or the nervous system overtaxed, or the function of an organ increased, we find an increase in the disintegrated products eliminated from the system, which shows that more waste has taken place than when there is less activity. As no function in the physical economy can be performed without life existing, and, as life cannot exist without vital energy, the cause of disintegration, which is death, must be the relaxing of the vital or life force. The moment the vital force ceases its control over the molecular or physical force, disintegration ensues. To replace this waste with living structure, the vital energy again controls the physical forces; if this did not take place, and disintegration continued, the system would soon be reduced to a condition where life could not be sustained. In some parts of the organism, the waste is more rapid than in others; this occurs where the vital energy is the most active. Waste and growth vary with age; from infancy to manhood or womanhood, the the growth exceeds the waste; from the latter to the decline of life, they are nearly balanced. In old age, the waste exceeds the growth and at death growth ceases; the vital energy is suspended and no longer controls the molecular or physical forces, and disintegration of the body follows as a natural result. In proof of the vital energy being the master-workman: whenever it is overtaxed, we find the physical organism giving way, the waste being greater than the growth, the system soon becomes dilapidated and the vital force is gradually compelled to yield to the molecular, which, if continued, must terminate in death. We have another proof of the control of the vital force over the physical in the unfecundated egg. If it be submitted to a normal temperature, it will decompose; an egg fecundated, submitted to the same temperature, will develop into a chick. Many more facts could be given to verify our assertion, that the life-action taking place in the physical organism does not result from chemical or physical change, as Herbert Spencer and others would have us believe. His definition of life-action is: "A definite combination of heterogeneous changes, both simultaneous and successive in correspondence with external co-existences and sequences." Or, as G. H. Lewis defines it, "A series of definite and successive changes, both of structure and composition, which takes place within an individual without destroying its identity." Neither attempts to define what it is that causes these changes and sustains the organism through physical life.

The reader will naturally inquire what constitutes the vital force. We have already discussed what it is in Chapter II., when speaking of the function of the nervous system. It is the energy of the soul, which unfolds from the cell-germ with the physical body, and is the force imparted by the parents. It is therefore purely a subjective force, and correlates with the organic nerve forces, the two combined constituting the nervo-vital forces. Growth is an integration, a building up of organic structure, in which the vital and nerve forces are correlated. In disintegration, or decay, the correlation is suspended, and the physical forces are permitted to have control, reducing the tissues to their normal chemical condition. Death is nothing more nor less than a general disintegration. Nutrition is a nervo-vital action, in which the subjective or vital force, and the objective or nerve force, both participate, being correlated. When nutrition is impaired, as it is in Bright's Disease, the correlation of the two forces is disturbed. When a person is in vigorous health, and has what is usually termed a robust constitution, the vital energy is active; in impaired health and in old age it is inactive. We will therefore perceive that instead of life being the result of chemical or molecular changes, it is the reverse—it is

A MALIGNANT GROWTH. 103

death, not life. The scientists have placed the cart before the horse; they have Nature turned up side down, or round about.

If the nervous system undergoes the slightest structural change from impaired nutrition, or other causes, the correlation of the nerve and vital forces is impaired to the extent of the pathological change. Every intelligent physician knows that serious functional disturbance may take place in the nervous system, as in convulsions, lock-jaw, etc., causing death, without leaving the slightest structural change behind. In such cases there is no derangement of the nervous structure; it is merely an imperfect correlation of the forces. Such violent functional derangements may be compared to an organic cyclone or tempest, which is difficult to overcome, owing to the impossibility of re-establishing the correlation of the forces.

About three years ago, the author concluded, after a strong appeal being made to save a life, to test his views of life-action in the treatment of a malignant growth. The unfortunate individual was a gentleman in his seventy-third year. He had a large fungus hæmatodes on his arm, just below the shoulder; an eminent surgeon was called in to examine the case with a view of removing it, or amputating the arm in order to get rid of the stench, if nothing more. He declined, stating that an operation could not be borne, and even if it could, it would not prolong life a great while, for his system was "*rotton with cancer.*" I undertook the case with no great expectation of benefiting him. In about three weeks after commencing treatment, a shrinkage was observed with the lessening of the discharge and odor. A few days after, the outer border of the tumor appeared to deaden and drop off, leaving a healthy surface. This continued until the whole abnormal growth had disappeared, and the parts healed. The time occupied in the treatment was about four months. There were no local applications used, the remedies were given with a view to improve nutrition, and to change the growth from an abnormol to a normal one. It is over two years since this case was treated and there has been no return. A similar plan of treatment must be adopted in Bright's Disease, in which there are two conditions that must be met: one is the congestion of the kidneys and the other a structural change taking place in the cortical portion of the same. In the former, the capillaries surrounding the uriniferous tubes are relaxed, owing to an ennervation of the vaso-motor energy permitting too much blood to be eliminated from the capillaries for the formation of a normal basement membrane, causing an increased cell development, and a corresponding disintegration. It is well known among physiologists that the activity of epithelium cells depends upon the amount of blood supplied and the nervo-vital energy imparted to it. If the capillaries of the salivary glands be dilated, the increased flow of blood to the glands causes a rapid increase in the secretions, and a more rapid disintegration of the cells; on the contrary, if the blood-supply be diminished, the secretions are lessened and become more fluid or watery. If the nerves leading to these glands be divided, the secretions become vitiated or are entirely suspended, proving that nervo-vital energy and blood-supply, are the essentials for cell-action. The same will apply to all secretory or excretory cell structures. If the secretions are increased, there must be a corresponding increase of cell formation, and a corresponding cell disintegration.

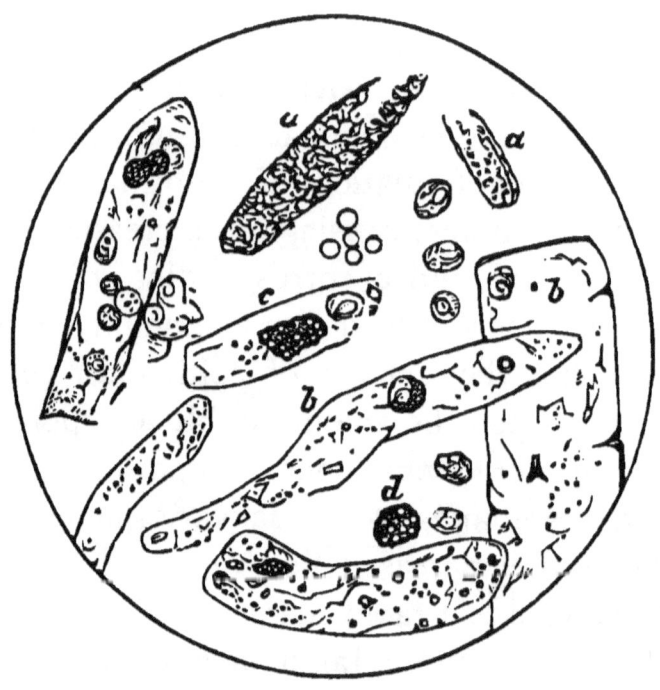

FIG. 9.
Morbid products observed in the urine in Bright's Disease: *a*, epithelium casts; *b, b*, transparent or hyaline casts; *c, d*, fatty matter. (From Reynolds.)

This condition we find in the kidneys in Bright's Disease. In its early stage, there is a congestion of the kidneys from a relaxation of the capillaries, permitting an increased elimination of blood serum to the

GROWTH AND DISINTEGRATION. 107

basement membrane, an increase of epithelium cell formation, and an increased secretion, providing the nervo-vital energy is not too greatly ennervated. As activity

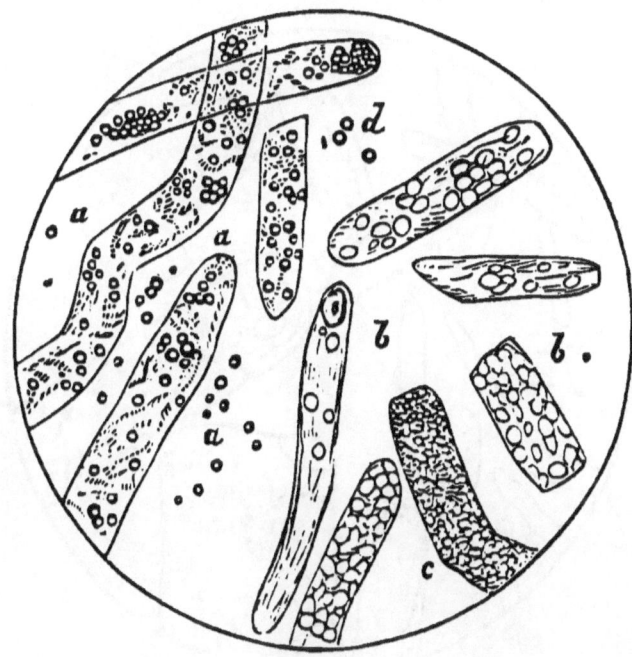

FIG. 10.
Morbid products eliminated by the kidneys in Bright's Disease, and observed in the urine by the aid of the microscope. *a, a,* fatty casts; *b, b* granular casts; *d, d,* free fatty molecules. (From Reynolds).

of growth must be attended by a corresponding disintegration, we find a large amount of epithelium cells in the urine. As the nervo-vital energy becomes more exhausted, we find the cell-lining of the tubes thrown off before they have accomplished their office, which is the elimination of the solid contents of the urine. These are called epithelium tube-casts [Figs. 9,

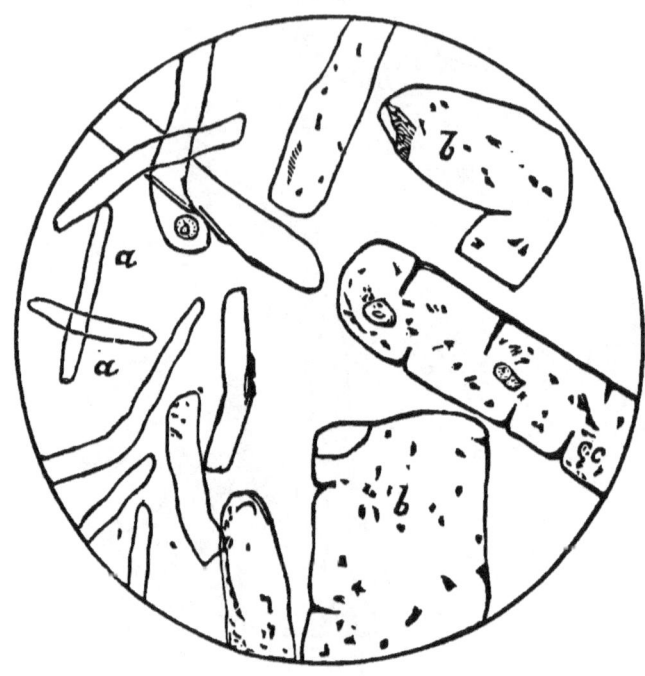

FIG. 11.

a, a, transparent casts observed in the urine in a case of chronic Bright's Disease of eight months' duration; *b, b*, the same form of casts from chronic Bright's Disease (large white kidney); *c*, the same form of casts from a case of chronic Bright's Disease (contracted kidney, with fatty degeneration). (After Reynolds.)

10 and 11]. As the exhaustion continues, the cells are not formed, there not being sufficient vital energy imparted to the substance of the basement membrane to

develop the growth of cells beyond the granular layer [Fig. 5].

If a still greater exhaustion of nervo-vital energy occurs, we find but slight traces of vital action in the basement membrane; no more than is sufficient to organize the membrane into a semi-gelatinous condition, as we find in the amœba, or lowest organisms. In this stage of the disease we find in some of the tubes no trace of the granular or epithelium layers, nothing more than a structureless membrane, which is thrown off and called transparent tube-casts [Fig. 11]. The formation and elimination of the three forms of tube-casts have never been explained. The views we have advanced of vital energy and life action indicate the progress of Bright's Disease as it is manifested in the kidneys. With the aid of the microscope we may, by observing these different tube-casts in the urine, ascertain the progress of the disease by the too rapid disintegration, which is not checked because the vital energy has become weakened. While these changes in the cortical substances are going on, the medullary portion of the kidneys becomes involved, owing to the same ennervation.

In order to impress upon the mind of the reader, we assert that no life action can take place in the material organism without the correlation of the soul or vital force and molecular nerve force, the action of the two constituting physical life. Simple hypertrophy is a derangement of the correlation of the nervo-vital energy, with an *increase* of the vital over the molecular energy, inducing change of form, but not of type. Simple atrophy is a derangement of the correlation of the same energies, with a *decrease* of the vital, also causing change of form, but not of type. When the change of type and form both occur we have abnormal growth, the malignancy of which will depend upon the extent of the impaired correlation. In the most malignant form of cancer, such as the fungus hæmatodes, the correlation of the forces is almost suspended, and the vitality of the part is at a very low ebb. The same law will apply to all structural changes, or functional derangements occurring in the living organism.

It would afford us pleasure to dwell

more fully on this important and interesting subject of vital energy, but space will not permit, and we shall therefore refer the reader, if interested in the subject of life, and how it may be prolonged, to a subsequent work we hope soon to have ready for the press.

Before closing the chapter we wish briefly to express our disapprobation of the stress laid on pathological changes observed after death, as a chief guide to diagnosis of diseased action. It is remarkable with what care physicians, in examining bodies after death, note the color, form, consistency, etc., of the diseased parts, without searching for the cause of their abnormal changes. The heart may be as large as that of an ox, or as small as that of a dog; the brain as hard as a Dutch cheese, or as soft as lard; the liver as white as the face of a Caucasian, or as dark as the face of a Malay; the kidneys may be enormously enlarged, or shrunken to half their natural size. Yet what does all this prove? Simply a change of structure from a cause unperceived by the pathologists. Mr. Lee quotes in his "Mineral Springs of England," the following from the pen of Dr. Kreysig, which is pertinent to the question we are now discussing. He says: "Physicians are in the habit of regarding the solid parts as the primary agents of life, to which the fluids are subordinate; but, on the contrary, the blood and the *nervous system* are the primitive and essential instruments of all the organic functions, while the solid parts occupy the inferior grades, and are of but secondary importance in disease. The elements of general and internal disease, or the morbid predispositions which form the most important objects of treatment, may, then, all be reduced to a vitiated state of the blood and the lymph, or to the derangement of the nervous system." We would remark that the "blood" and the "lymph" could not become vitiated excepting through ennervation of energy of the organic nervous system. Professor George B. Wood* taking exception to the generally received opinion as to the cause of albuminuria, states: "The most probable explanation appears to be that there is some failure in the vigor of the assimilating process, dependent on

---

*Treatise on the Practice of Medicine, Sec. Ed., Vol. II, p. 508.

the same cause which has predisposed the
kidneys to the morbid deposition going on
within them, and that the two effects, instead
of having any mutual dependence, are
merely co-existing results of the same concealed
agency." Dr. Henry Hartshorn, in
the American edition of Reynold's System
of Medicine, Vol. III, page 662,* states that
"Semmola insists that true Bright's Disease
is always connected with; or consists
in a *general nutritive disorder*, to which the
affection of the kidneys is *secondary*. This
disorder begins, according to his statement,
in an arrest of *cutaneous respiration*. Next
follows imperfect digestion, and transformation
of albuminoid food; and the presence
of an excess of albumen, with deficient
formation of urea in the blood. Hence results
renal irritation and inflammation, albuminuria,
and diminished excretion of urea.
At a later period, when the kidneys have
undergone considerable morbid alteration,
the secreting structure fails to eliminate the
urea which is formed. Then, and not before,
will uræmia take place. Semmola con-

---

* *Gazette Medicale de Paris*, 1875, quoted in *British Medical Journal*, September 27th, 1879.

siders that pathologists have generally too much neglected the *dyscrasic* character of Bright's Disease, and its special dependence (being thus set apart by him from albuminuria, or other causes, *e.g.* alcoholism, gout, etc.), upon the gradual action of cold upon the skin." The source of Bright's Disease, from observation and treatment, is not to be found in the pathological changes observed in the kidneys after death. They are merely the effect of a cause that must be sought for and ascertained before it can be successfully treated. The disease may exist for a long time before any morbid change is observed in the functions and structure of the kidneys. Of this we shall speak more fully in the chapter on the cause of Bright's Disease.

As this work is not written for the profession, but for those predisposed to and suffering from Bright's Disease, we have not considered it advisable to give treatment, as any educated physician, who understands the functions of the organic nervous system, will be able to select suitable remedies to correct ennervation, overcome renal congestion, and improve nutrition. This is all that is required of the physician. Nature will accomplish the rest, providing too serious structural changes have not occurred in the kidneys, and too serious complications have not taken place in other organs. Physicians familiar with Bright's Disease must have observed that in many cases of the chronic form, before the disease has become established, serious local troubles have occurred in other parts of the system, This proves that the ennervation is not confined to the kidneys, but is very generally diffused, and by correcting it nutrition is re-established, structural changes gradually disappear, abnormal functions are restored to their natural condition, and health is re-established. The nervo-vital energy has great recuperative powers, and all that is required is assistance from a skilful and educated physician.

CHAPTER VI.

SYMPTOMS OF BRIGHT'S DISEASE.

WE have already stated that Bright's Disease is usually divided into acute and chronic, and the latter is again divided into three or more forms, corresponding to the pathological changes observed.

1. THE ACUTE FORM.—This form of the disease makes its appearance suddenly, and generally attended with chilliness, fever, headache, nausea and vomiting. There is frequently a dull, aching-pain in the region of the kidneys, with soreness on pressure, and skin dry. Dropsy usually appears very early and generally commences in the face, causing a fullness under the eyes, with a stupid or heavy expression of countenance. It soon extends to the upper extremities, making its first appearance in the hands. Soon the chest becomes involved, and, finally, the lower extremities. In males, the scrotum is a favorable location for the dropsy to make its appearance; sometimes it becomes almost as large as a child's head.

In a few cases, the swelling commences in the feet and extends upwards, becoming general. If the patient should be kept in a recumbent position, the back is the part most swollen; as the dropsical effusion increases, there is a corresponding lessening of the quantity of urine secreted. In some aggravated cases, it is completely suppressed, which is a very alarming symptom. In the early stage of the disease, the urine, instead of having the natural amber color, presents a smoky appearance, which is owing to the coloring matter of the blood. The urine has an acid reaction. and a specific gravity ranging from 1025 to 1030, and has been known to reach as high as 1060. If the disease passes into the chronic form, the specific gravity lessens and may fall as low as 1010. If the urine be permitted to stand for a short time, a flocculent, brownish sediment forms. If it be examined with a microscope, it will be found to consist of blood corpuscles, epithelial cells, free nuclei, coagulated fibrin and tube-casts, which consist of three forms corresponding to the three layers of the walls of the uriniferous tubes. They are described as the epithelial, the granular and the hyaline or transparent tube-casts [see Figs. 9, 10 and 11]. The first, called the epithelium, constitutes the secreting portion of the tubes; the second, the granular, lies immediately below the former, and the third is the hyaline or transparent, formed of lymph eliminated from the capillaries, of a semi-fluid character, and forming what is called the basement membrane of the tubes. When these layers are all thrown off, the tubes are obliterated or completely destroyed, leaving the capillary network, which is highly congested. Fortunately, it is only a part of the uriniferous tubes that is destroyed (excepting in extreme cases), the others being comparatively healthy, or merely stripped of a portion of their epithelium. There are also found in the sediment oil-globules, which are a normal product of the kidneys, found distributed among the tubules.

If the urine be boiled it will be found to contain a large amount of albumen which is the characteristic feature of Bright's Disease. The normal constituents of the urine such as urea, uric acid, etc., are very much reduced in quantity, owing to the structural change taking place in the uriniferous tubes, they being the natural channel through which they are eliminated. The drain of albumen from the blood impoverishes it and the retention of the nitrogenous products poisons it. The two combined depress the vital energies of the system, causing an imperfect assimilation which is shown in the diminution of blood corpuscles. The reduction of the red corpuscles of the blood and the anæmia caused by it, are attributed to the amount of blood eliminated by the kidneys. This is so small that it will not account for it; another cause must be sought for. We attribute it to imperfect assimilation caused by an ennervation of the organic nervous system.

The acute form of the disease may continue for a few days or many months, and the patient apparently recover, or it may pass into the chronic form and continue for many years. If it should have a favorable termination, the quantity of urine will be increased, the specific gravity diminished, the albumen will gradually subside and the tube-casts disappear in the urine. The skin becomes moist and the dropsical effusion absorbed. Convalescence is generally slow, and requires from two to six months for the kidneys to assume their natural functions and the system to recover from the shock. Recovery is not complete until the urine is entirely free from albumen, and the dropsical symptoms have entirely disappeared. If albumen continues in the urine with tube-casts, it is evidence of structural change going on in the cortical substance and of the disease assuming a chronic form. Sometimes in the acute form the albumen almost disappears from the urine, which may be merely a lull in the disease to be followed by an exacerbation more severe than the previous attack. A relapse of this kind may generally be attributed to cold and dampness, diseased kidneys being very susceptible to both.

The unfavorable symptoms of the acute form are an increase of the albumen, a diminution of the quantity of urine passed and a change in the color of the urine, which becomes darker, owing to the elimination of blood.

This form of Bright's Disease is occasionally accompanied with complications which are considered by some writers as local symptoms. For example, œdema of the lungs they regard as a general dropsy. Inflammation of the serous membrane very frequently takes place. Pleurisy is the most common; pericorditis is less frequent, and peritonitis very seldom occurs. Pneumonia is occasionally a complication but not so frequent as bronchitis. Hypertrophy of the heart rarely occurs in the acute, but is a frequent complication in the chronic, form of the disease.

cute Bright's Disease is not confined to any particular age; a large majority of cases occur during childhood and youth, and as a general rule the person attacked has previously enjoyed good health.

II. CHRONIC FORM.—The general symptoms of Chronic Bright's Disease cannot be more clearly given than in the following ex- extract from Reynolds' "System of Medicine," American Edition by Dr. Hartshorn:

"In the great majority of instances chronic Bright's Disease begins slowly and imperceptibly. The attention of the patient is awakened, some months, or it may be years, after it has existed, by the gradual failure of his strength and his increasing pallor or sallowuess, with a disinclination to exertion; or his suspicions are aroused by a little puffiness under the eyes, a slight swelling of the ankles at night, unusually frequent calls to void urine, or shortness of breath. In other cases these premonitions pass unheeded or are altogether wanting. The disease proceeds silently, amid apparent health, and then suddenly declares itself by a fit of convulsions, rapid coma, amaurosis, pulmonary œdema, or a violent inflammation; or it may lie concealed for an undetermined period, and then reveal itself, after exposure to cold or a fit of intoxication in the guise of an acute attack, with rapid general anasarca, scanty, sanguineous urine; or it may be a continuation or sequela of acute Bright's Disease; or lastly, it may creep on stealthily in the wake of some pre-existing chronic disorder, phthisis, caries, necrosis, constitutional syphilis, gout, chronic alcoholism, or exhausting suppuration.

"The principal symptoms of the disease are: albuminous urine with deposits of tube-casts and renal epithelium, especially at

GENERAL SYMPTOMS. 123

night; dropsical effusions into the subcutaneous cellular tissue, serous cavities, or pulmonary substance; dryness of the skin, derangements of digestion; progressive hydræmia; uræmic phenomena (headache, amblyopia, convulsions, coma, vomiting, and diarrhœa); hypertrophy of the left ventricle; secondary inflammation of the parenchymatous organs and serous membranes.

"Few cases present the whole of these symptoms: many present only two or three of them. The alterations in the composition of the urine are the most invariable, and also the earliest and most distinct symptoms; next follow in the order of constancy, the deterioration of the blood, the dropsical symptoms, and, lastly, the uræmic and inflammatory incidents.

"The disease usually pursues an interrupted course, being subject to occasional exacerbations, with intervals of quiescence. The exacerbations are generally by exposure to cold, or some imprudence in diet or regimen; sometimes no cause can be assigned for their occurrence. They are marked by pyrexia, and often simulate an attack of acute Bright's Disease. The intervals of quiescence may be some weeks or months, or a few years; the remission of the symptoms is commonly only partial, the main features of the disease persisting, though in a modified degree. Sometimes, however, the remission is almost complete, and little except the albuminous state of the urine remains to attest the existence of renal mischief; and even this may, in exceptional cases, be absent, and the nature of the case be first revealed at the autopsy. After each exacerbation it is commonly pretty evident that the disease has progressed a step, and probably an additional portion of the kidney, hitherto spared or only slightly affected, has been disabled. The kidneys are at length so seriously disorganized, and their depurative functions so far abrogated, that life becomes impossible.

"The immediate cause of death is variable. Sometimes the sufferer passes quietly away, exhausted by anæmic, burdensome anasarca, and defective digestion. About one-third of the subjects of chronic Bright's Disease perish by uræmic poisoning, either in the form of coma and convulsions or irrepressible diarrhœa and vomiting. A considerable number die from the intensity or dangerous situation of the dropsical effusion, as when the glottis or lungs are invaded; or death results from the hydrothorax, or from gangrenous erysipelas set up in the tense œdematous integuments of the legs, thighs, or genitals. About one-fifth die by secondary pneumonia, pericarditis, or double pleurisy. The remainder are cut off by remote complications, as apoplexy, cirrhosis, phthisis, intestinal ulcerations, etc.

"The *duration* of the disease can only be approximately ascertained from the difficulty of assigning the exact date of invasion. Enough is, however, known to show that it varies within very wide limits. The usual period is from two to three years; but cases may end in six months, or be protracted for four or five years. Exceptional instances have been recorded in which patients have survived ten years, and even fifteen and twenty years."

SPECIAL SYMPTOMS AND COMPLICATIONS: URINE.—"The quantity of albumen is variable. The urine may become absolutely solid on boiling, or it may contain only the minutest traces of albumen, even in confirmed and fatally-tending cases. The amount of albumen lost in twenty-four hours varies commonly from forty-five to three hundred grains; Dr. Parks observed in one instance, five hundred and forty-five grains. During digestion, the quantity is larger (it may be double) than during fasting; it rises and falls irregularly in the course of the disease, sometimes diminishing to a trace, and anon increasing to an intense impregnation.

"The urine is generally pale and slightly turbid, depositing, on standing, amorphous whitish sediment of renal epithelium and tube-casts. It sometimes contains blood, occasionally in quantity, but generally in microscopic proportions. . . . . When the case is complicated with phthisis, or regurgitant heart disease, the urine may be high-colored and turbid from lithates.

"The quantity of urine voided per day varies with the type of the disease, and the presence or absence of pyrexia, sweating, vomiting, or diarrhœa. The specific gravity is low when the urine is copious (1006 to 1010); but when scanty, the specific gravity may rise to 1030, or even to 1040.

The urine is nearly always acid, and not unfrequently deposits uric acid and oxalate of lime. Occasionally I have noted alkaline from a fixed alkali, and on two occasions ammoniacal on emission. The *renal derivatives* are markedly scantier in chronic than in acute Bright's Disease; they are not unusually entirely absent for limited periods. They are, however, sometimes discoverable when the urine has temporarily ceased to be albuminous. The epithelium cells may be simply withered; more rarely they are totally disintegrated into an amorphous granular debris; in other cases they contain specks of oil, or they may even be wholly converted into an agglomeration of oily particles, so as to appear identical with the granular corpuscle, or inflammation globule. If the casts are similarly speckled with fat, and free oily dots are scattered over the field, it indicates a fatal disorganization of the organs—either large, fatty kidneys, or contracted granular ones. Considerable diversity in the character of the casts, discharged by the same individual, even during the same day, may be met with, arising from a different condition of the several parts of the gland; conclusions as to the probable state of the kidney can only be drawn from the *prevailing* character of the deposit, and not from one or two individual casts or cells. The casts most commonly seen in chronic Bright's Disease are 'small' and 'large' hyaline forms and 'granular' opaque ones, any of which may have a few wasted epithelium cells strewed over them. Perfect 'epithelial' casts are rare; blood-casts are also rare, in chronic cases, unless there be concomitant tricuspid regurgitation. Large hyaline casts result from exudation having been thrown into tubuli denuded of their epithelium. [The more rational conclusion is that the hyaline cast is the basement membrane that has not advanced to the granular state, and that the granular cast is the basement membrane that has not advanced to the epithelium state. S. P.] The longer the exudation is retained within the tubuli, the darker and more granular the casts, and *vice versa*; casts speedily discharged are commonly hyaline. Sometimes casts are darkened by the coloring matter of the blood, and the opaque granular ones are sometimes composed of crushed epithelial *debris*, moulded in the form of tubuli. [Are not the opaque granular casts the granular base with imperfectly formed epithelial cells? S. P.]

"The normal solids of the urine are all diminished in chronic Bright's Disease. The urea is, as a rule, markedly reduced, the daily quantity averaging only about one hundred grains; Frerichs has observed it as low as fifteen grains. A case is mentioned by Mosler, however, in which six hundred and forty grains were voided in one day! There is no correspondence, direct or inverse, between the urea-excretion and the discharge of albumen."

BLOOD.—" The changes in the blood are the complement of those in the urine; it becomes more watery and poorer in albumen and red corpuscles, while the urea, uric acid, extractive matter, and pale corpuscles are relatively increased. This alteration in the composition of the blood is deeply concerned in the production of the more prominent features of the disease—the anæmia, dropsical effusions, uræmic phenomena, and secondary inflammations.

" DROPSY is much oftener absent in the

chronic than in the acute form. It is much more constant with the smooth large than with the granular contracted kidney, of which probably one-third or one-fourth of the cases run their entire course without dropsy. The effusion begins quite as often in the feet and legs as in the face, and is apt to change its seat capriciously; sometimes it is excessive and general, but usually slight and partial. When the heart or liver is diseased, ascites and œdema of the legs become unduly prominent. The effusion may disappear totally for months, and then return; more frequently after the subsidence of the general dropsy, œdema lingers obstinately in one or more places, over the flat of the tibiæ, about the ankles, beneath the eyelids, or about the genitals. The presence or absence of dropsy generally, but by no means always, corresponds with the abundance or scantiness of the urine, but it has no relation to the amount of albumen.

"THE SKIN is usually obstinately dry; perspiration is quite exceptional. . . . . . Profuse sweating does, however, sometimes takes place spontaneously, and may even continue for weeks. . . . . The integuments in some cases are excessively pale and glossy, more commonly sallow and rough. There is little or no tenderness in the renal region in the chronic cases, and the frequency of micturation is mostly observed at night. Some degree of bronchitis is almost an invariable coincident in both acute and chronic Bright's Disease.

"COMPLICATION AND CONNECTION WITH OTHER DISEASES.—The digestive organs are nearly always disturbed; at first, there is anorexia [loss of appetite] and nausea, and, later, frequent or even uncontrollable diarrhœa and vomiting are not unusual. . . . Secondary inflammation of the lungs, endocardium, pericardium, peritoneum, or integuments, may break out at any period in the course of Bright's Disease, and the tendency to these constitutes one of the principal dangers of the complaint.

"BRIGHT'S DISEASE AND PHTHISIS.—This complication is of frequent occurrence. In the great majority of cases the pulmonary disease is far advanced before renal symptoms appear, the long-continued discharge of pus from the lungs at length giving rise to waxy changes in the kidneys, followed by albuminuria and dropsical effusions. But sometimes the renal disease precedes the pulmonary, and the changes found in the kidneys after death are not invariably of the waxy type."

Hypertrophy of the heart, especially of the left ventricle, is a very frequent attendant in chronic Bright's Disease, which is supposed by Traube to be caused by increased tension in the arterial system, producing an increased resistance, which the left ventricle has to overcome.

URÆMIA.—The prominent symptoms are drowsiness, impaired vision, and dullness of hearing, nausea, vomiting, diarrhœa, and finally coma or convulsions. The cause is the retention in the blood of disintegrated products, owing to a derangement of the functions of the kidneys. The headache is generally in the forehead, vertex, and back of the head, and deep in the orbits. There is also a sense of weight and compression, attended with dimness of vision, which comes and goes, or may be attended with temporary loss of sight. The coma comes on insidiously, and may not be noticed by those in attendance, supposing it to be nothing more than drowsiness, until convulsions occur. Not unfrequently the coma and convulsions occur simultaneously. As a rule the urine is very greatly diminished, or completely suspended before the coma and convulsions take place. These may be regarded as alarming symptoms, and if not relieved promptly, will terminate fatally within a short time.

It will not be amiss to mention an obscure symptom of Bright's Disease, which, if not interesting to the general reader, may be of service to physicians, if they have not already observed it. It is a feebleness of the heart, causing a weak and compressible pulse, which becomes more so as the disease advances. The heart's energy is derived from the medulla oblongata, through the pneumo-gastric nerve. The medulla is the co-ordinating centre of the spinal system, and the solar ganglion the co-ordinating centre of the organic or sympathetic nervous system. The pneumo-gastric nerve distributes the energy that controls respiration, and the action of the heart and its branches anastomose with the solar plexus. It is from this plexus that the uriniferous tubes (the seat of Bright's Disease in the kidneys) receive their nerves. It is well known among physiologists that an injury to the medulla, or to the pneumo-gastric, brings about albuminous deposits in the kidneys, as well as diabetes. This shows the intimate relation existing between the two great co-ordinating centres, the medulla oblongata, situated at the upper portion of the spinal cord, and the semi-lunar ganglion, the centre of the organic nervous system.

I would also remark that one of the first evidences of improvement in Bright's Disease is in the action of the heart giving more tone to the pulse, thus enabling it to offer more resistance to pressure. The enfeebled condition of the pulse is not confined to Bright's Disease, but is frequently found in various forms of chronic diseases. In all cases exhibiting this form of pulse, with symptoms favoring Bright's Disease, I have invariably examined the urine, and very frequently found traces of the disease. It therefore gives us a *cue* in many cases which might otherwise be overlooked until the disease becomes permanently established.

CHAPTER VII.

CAUSES OF BRIGHT'S DISEASE.

.THE causes of disease may be classified as primary and secondary, the latter recognized as the exciting cause. The primary is an ennervation of the organic nervous system, the secondary is the one that brings out, or makes manifest, the primary cause. As Bright's Disease assumes two distinct forms, *acute* and *chronic*, and the exciting causes being different in the former from what they are in the latter, we will, in order to make the subject clear, speak of them separately.

1. SECONDARY, OR EXCITING CAUSES OF THE ACUTE FORM.—A large percentage of the acute cases of Bright's Disease are sequelæ of scarlet fever, and other infectious diseases, such as diphtheria, small-pox and measles; skin diseases and extensive burns are also recognized as secondary causes. Some of the mineral and vegetable poisons are known to have been the means of developing Bright's Disease by causing

inflammation of the kidneys. Among the mineral poisons are arsenic, lead, nitrate of silver, and mercury, and among the vegetable, turpentine, phosphorus, oils of wormwood and mustard. Cantharides, which we find in some medical works classified with the vegetable poisons, is not a vegetable, but an animal product; it is a very active, exciting cause, inflaming the kidneys, and, in fact, the whole urinary tract. Alcoholism is a very frequent cause of both the acute and chronic forms. What is called "*moderate drinking*" is potentially an exciting cause not only of Bright's Disease, but of gout, heart troubles and liver complaint. Physicians are almost universally of the opinion that the drinking of alcohol in any form, save at meals, should not be indulged in. Pregnancy is also a cause, and generally occurs in lymphatic persons of a weak or feeble constitution with soft and flabby muscular fibre. Protracted and low forms of fever are occasionally a cause of the acute form of this disease.

2. SECONDARY OR EXCITING CAUSES OF THE CHRONIC FORM.—The causes of this form of Bright's Disease are more numer-

ous and more insidious in their action than those of the acute. Perhaps the most frequent causes are fast living, mental strain, anxiety and moderate drinking. Overindulgence in eating or drinking impairs the functions of the digestive organs, particularly the stomach and liver. The kidneys are overtaxed in the elimination of the superabundance of nitrogenous products of the food not required for the growth of tissue, and which must be eliminated. The liver being overtaxed from the excessive carbon and hydrogen of the alcoholic drinks, becomes weakened in its functions, and the kidneys are called upon to eliminate not only the superabundance of food, but also the carbon and hydrogen that the liver cannot. In this way the kidneys become involved.

We can thus account for the frequency of the disease in apparently strong and vigorous constitutions, in persons of fine physique. Among statesmen and the mercantile community the active cause is mental strain, sleeplessness, and want of proper rest—in many cases accompanied with anxiety. During the war it became more frequent and after the collapse of business in 1875 it increased. Alcoholism without anxiety or mental strain is sufficient to cause Bright's Disease; a confirmed tippler whose system is saturated with alcohol is very apt to have fatty degeneration of the liver and kidneys. Frequent exposure to cold and dampness by checking perspiration often causes congestion of the kidneys and leads to a morbid condition which is subsequently increased by chilling of the body.

Sexual excesses, consumption, syphilis, deep-seated abscesses, long-continued suppurating surfaces, are also active causes. Malaria of late years is assigned as a cause by depressing the vital energy of the organic nervous system. Heredity is another cause, which in one or more subsequent generations, owing to the present frequency of the disease, will very likely show itself actively. The weakness of parents is imparted to their offspring, and will crop out at some period of life; if not in their children, it will in their grandchildren. It is not an unfrequent occurrence for a hereditary disease to skip a generation; when it does, it generally reappears in an aggravated form. Simon found that confining animals in a dark room for a long time produced a fatty degeneration of the kidneys. Was it the absence of light, or the fear and despondency induced by the confinement? Light is evidently a vital tonic, but it is questionable whether the simple confinement in darkness would be sufficient to produce structural change in the kidneys.

Chronic Bright's Disease more frequently occurs between the ages of 30 and 60 years, seldom in children. Dr. Dickinson observed the disease in a lad of five years and Dr. Gee in the case of a boy of two and a half years.

The following table of 61 cases of the lardaceous form has been compiled by Dr. Dickinson:

Age	No. of Cases
0–10	3
11–20	11
21–30	21
31–40	10
41–50	10
51–60	3
61–70	3
Over 70	0

Dr. Dickinson also gives a table of 308 cases of the granular kidney.

1	was between	11 and 20
24	were	21 and 30
50		31 and 40
93		41 and 50
76		51 and 60
47		61 and 70
17	Over	70

3. THE PRIMARY CAUSE.—*Whatever cause will produce inflammation of the kidneys, when there is an ennervation of the organic nervous system may produce Bright's Disease.* The same may be applied to the lungs in consumption. *The ennervation is the predisposing or primary cause and an active or exciting cause makes it manifest.*

A predisposing cause, whether it be hereditary or acquired, may lie dormant in the system for years, as we frequently find in consumption, scrofula, Bright's Disease, etc., and suddenly be made manifest by a cold, or some other exciting cause. If physicians, particularly those who make a specialty of chronic diseases, were to seek more earnestly for the cause, they would find their efforts crowned with greater success than they now are. To be successful in the treatment of diseases, the primary cause must be sought for and removed before functional and structural changes, which are only the effects, can be checked.

CHAPTER VIII.

Advice to Those Suffering From Bright's Disease.

In the acute form there is but little advice to give, as disease is ushered in so suddenly, and in many cases symptoms are so urgent as to require at once the services of a physician. His first efforts will be to relieve the kidneys; he will treat other symptoms as the case demands. In the chronic form it is different. The disease manifests itself insidiously, and progresses very slowly. The kidneys are morbidly susceptible to cold and dampness, from which they must be protected; hence warm clothing is indispensable. and must be of such a character as to prevent the body from becoming chilled, which checks perspiration. Flannel should be worn the entire year. During the winter months the garments should be heavy, and consist principally of wool. Cotton flannels are of very little use; they do not prevent the elimination of caloric to any great extent. During the summer months lighter flannels should be worn, to be changed for heavier when there is a marked fall of temperature. Many patients forget, or do not consider it important, to change clothing to meet the changes of the weather, particularly during the spring and fall months. I have for years recommended light raw-silk shirts worn under the flannel. During the summer months they should be worn without the flannel. In Consumption and chronic Bright's Disease I frequently recommend silk shirts, drawers and socks, with the happiest results; the feet should be well protected from dampness by wearing double-soled shoes, with gum cloth or hog's bladder between the soles, to prevent the dampness from striking through. If they are double-uppers the hog's bladder may also be placed between them. Gum shoes may be worn during wet and muddy weather, but should be removed immediately upon entering the house. The least dampness of the feet will often cause the kidneys to become suddenly congested.

Alcoholic drinks should be avoided, unless they are really necessary to assist in supporting the system. In such cases, good Holland gin or Jamaica rum may be taken at meals. The mind should be free from care and anxiety, and moderate exercise is beneficial rather than otherwise, providing there is sufficient rest for nature to recuperate the brain. Cheerful spirits, a contented mind, accompanied with a clear conscience, and firm will, will do wonders in warding off diseases and prolonging life. A proper amount of sleep is very essential; it enables the brain to recover from the depression which ennervates the vital energy of the organic nervous system through the pneumo-gastric nerve. All those who have experienced the loss of sleep know how it depresses the system and how refreshing is a good night's rest. The sleeping apartment should be well ventilated by an open fire-place, grate, or some one of the modern means of ventilating bed-rooms. The lowering or raising of a window is objectionable for fear of a draught, particularly if the room be small and the bed placed near enough for the current of air to strike it.

The diet should be nutritious and of a mixed character; it is customary for physicians to discourage the eating of much animal food, on account of increasing the nitrogenous products of waste in the blood. Animal food is essential for healthy nutrition, but it may be advisable in the more advanced stage of the disease, when there is a suppression of the function of the kidneys, to dispense with it. But in the earlier stages I would recommend a good nutritious diet, consisting in ordinary proportions of animal food. In Bright's Disease the trouble lies in the ennervation of the organic nervous system. There is deficient assimilation of all the products of nutrition. Food is of no use unless it is properly vitalized; when taken into the system, it is dead matter and must be vitalized before it can supply the wants of the organism. This is accomplished during the passage of the chyle through the mesenteric glands and liver. If assimilation should be imperfect, vitalization is deficient, and it undergoes chemical change, the same as the disintegrated products after life-action has been expended, and is eliminated by the kidneys and other excretory organs in the same manner as products of waste. I therefore approve of a mixed diet of a nutritious character, providing the patient is treated with a view to increase the vital energy of the nervous system. As it is increased, animal food may be given in larger quantity once or twice a day, for breakfast and dinner. A light supper, consisting of bread and milk, oatmeal or crushed wheat and milk; this will keep the bowels regular, aid digestion and not interfere with sleep. I would also recommend, if there is much debility, milk taken at intervals of three or four hours, and, if the system seems to require it, I would suggest rum or gin punches taken two or three times a day. Exercise in the open air, either by walking or riding, is conducive to health, and, therefore, should be taken advantage of whenever opportunity affords.

Bathing is advisable; a tepid bath once or twice a week, followed by a good rubbing of the skin to increase its circulation, is very desirable. It relieves the congestion of the kidneys and increases the secretion of the skin, and thus assists in the elimination of urea and uric acid from tbe blood, which the kidneys are enabled to do only to a

BATHING, RUBBING, ETC.

limited extent. I have been in the habit of recommending to patients, in connection with bathing, the rubbing of the skin with a coarse crash towel or Turkish rubber, saturated with a solution of rock-salt of sufficient strength to bear an egg. When saturated, it is to be hung up to dry; the body and limbs are to be rubbed with this prepared towel once a day, morning or evening. After it has been in use one week, it is to be washed, dried and again saturated and used in the same way. This rubbing increases the circulation of the skin, and wonderfully relieves the kidneys of their congested condition. It accomplishes more, it increases the action of the secreting glands of the skin and eliminates the products of waste.

As regards treatment, I have but little to say, as this work is not written for physicians, and if the treatment were given, it could not be made use of without consulting a physician. Therefore, my advice is, to avoid quacks as a saint would the fallen angel, and place yourself in the hands of a skilful doctor. There is, however, one form of treatment that I must condemn, for I have frequently seen its injurious effects. I have reference to the administration of iron in enormous doses; "*saturating the system*" with it in order to overcome anæmia. This is a great mistake for several reasons. In the first place, it will not cure anæmia so long as the vital energy is so ennervated that it cannot carry on a normal assimilation. Anæmia results from an impaired vitalization of the chyle in its passage through the mesenteric glands, and therefore the white corpuscles cannot progress or advance to the red corpuscles or blood-discs. There is nearly always sufficient of iron in the food and drink we take into the system for the purposes of the economy. The vital energy of the nervous system has its own laboratory in which it combines the organic and inorganic constituents in sufficient quantities to suit its purposes, and it is during this combination that they are vitalized. If there should be an excess of any one or more ingredients, it is rejected and eliminated, the same as the disintegrated products are by the excretory organs. There is another objection to giving iron in massive doses, it deranges the stomach and irritates the urinary organs, particularly the bladder. If it is to be administered at all in Bright's Disease, which is questionable, it should be given in very small doses, for an exceedingly small quantity is required for assimilation. It will thus be seen that instead of being beneficial in large doses, it is highly injurious. There is another form of treatment which I must caution those suffering from this disease to shun as they would their bitterest enemy. I refer to the numerous remedies, advertised by quacks, for the cure of Bright's Disease. They are usually powerful diuretics which act upon inflamed and irritable kidneys, forcing their function, which they are unable to perform owing to impaired energy, and thus increasing the difficulty. I am not surprised at those afflicted with this disease seeking such a source for relief, for the regular physician gives the patient no encouragement, and he is like a drowning person clinging to a straw. The kidneys require gentle treatment and any remedy administered that will force them must be injurious.

Much more might be said on the subject of which this chapter treats, but as our space is limited we will conclude by advising the patient to avoid quacks and their medicines and seek the advice of an intelligent and skilful physician, and that this be done before the disease is too far advanced. Chronic Bright's Disease is curable, even after the kidneys are seriously disorganized, providing the inflammation is removed and the energy of the organic nervous system re-established. The master-workman—the vital force, located in the system, calls for help, and if proper assistance be rendered in time, it will be enabled to restore the parts to their normal condition and re-establish a healthy function.

ANNOUNCEMENT.

We desire to announce to the public that within three or four months we expect to have ready for the press, a work of about four hundred and fifty pages, 8vo., entitled "Life, its Nature, Source, Purpose and Destiny, and the Method and Means to Prolong Physical Life."

Scientists admit that life exists but they have hitherto been unable to give its source, and how it becomes manifested in organic forms. We shall endeavor to trace life from its source and declare the forces and laws by which it is enabled to unfold life-organisms and sustain them for a limited period, and show what becomes of them when physical life ceases. In doing so we shall trace energy and law from their primary source, and show how the life principle controls them in the unfoldment of organic forms.

This subject has been our almost constant study for over thirty years, and we expect to present it in such a manner as to be understood by the ordinary reader. The work will be illustrated with plates and diagrams which will aid the reader in understanding this interesting subject. When it is ready for the press a prospectus will be issued, giving an outline of the work and a description of the illustrations.

www.ingramcontent.com/pod-product-compliance
Lightning Source LLC
Chambersburg PA
CBHW030348170426
43202CB00010B/1290